AUTOMATED TELLER MACHINE
BASIC FUNDAMENTALS

AUTOMATED TELLER MACHINE
BASIC FUNDAMENTALS

by
Chinmoy Saha
Anil Samuel
Raja J

Wordpen Academics
Publishing Science That Matter

ISBN: 978-93-6084-190-4

Published by : **WORDPEN ACADEMICS**
4736/23, Ansari Road, Darya Ganj,
New Delhi – 110 002
Phone: +91-011-35004336
E-mail: info@wordpenacademics.com

Printed at : **Replika Press Pvt. Ltd.**

Exclusive Distributor: ***Bio-Green Books, New Delhi***

Preface

Automated Teller Machine: Basic Fundamentals is an introductory guide designed to provide a clear and comprehensive understanding of the technology, functionality, and operations behind Automated Teller Machines (ATMs). As a ubiquitous component of modern banking and finance, ATMs have revolutionized the way we manage and access money, offering unparalleled convenience and efficiency in financial transactions.

This book serves as a foundational resource for students, banking professionals, engineers, and anyone interested in the workings of ATMs. It covers the core principles of ATM design, technology, hardware components, software architecture, security protocols, and transaction processes, providing readers with both the theoretical knowledge and practical insights needed to understand ATM systems.

Throughout the chapters, we delve into the history and evolution of ATMs, exploring how they have become integral to financial institutions' operations and customer service. The book also covers key topics such as ATM communication networks, security challenges, fraud prevention, and emerging trends like biometric authentication and mobile integration. These aspects are essential for understanding how ATMs continue to evolve in response to the ever-changing demands of the digital economy.

The goal of this book is to demystify the complex technologies and systems that power ATMs and to highlight their critical role in the broader financial ecosystem. By offering a detailed yet accessible explanation of how ATMs operate, we aim to provide readers with the knowledge to engage with ATM technology effectively, whether in a technical, managerial, or consumer-facing context.

We would like to extend our gratitude to the professionals, engineers, and researchers whose contributions to the field of ATM technology have shaped this book's content. It is our hope that Automated Teller Machine: Basic Fundamentals will serve as a valuable resource for those looking to explore the essential role of ATMs in modern banking.

Chinmoy Saha

Anil Samuel

Raja J

Contents

Chapter 1
Introduction to Automated Teller Machine (ATM)

1.1. Overview

An automatic teller machine (ATM) is a computerized telecommunications device that provides the clients of a financial institution with access to financial transactions in a public space without the need for a cashier, human clerk or bank teller. On most modern ATMs, the customer is identified by inserting a plastic ATM card with a magnetic stripe or a plastic smartcard with a chip, that contains a unique card number and some security information, such as an expiration date or CVVC (CVV). Authentication is provided by the customer entering a personal identification number (PIN).

Fig ATM machine

Using an ATM, customers can access their bank accounts in order to make cash withdrawals (or credit card cash advances) and check their account balances as well as purchasing mobile cell phone prepaid credit. ATMs are known by various other names including automated transaction machine, automated banking machine, money machine, bank machine, cash machine, hole-in-the-wall, cashpoint, Bancomat (in various countries in Europe and Russia), Multibanco (after a registered trade mark, in Portugal), and Any Time Money (in India).

ATMs are placed not only near or inside the premises of banks, but also in locations such as shopping centers/malls, airports, grocery stores, petrol/gas stations, restaurants, or any place large numbers of people may gather.

Fig ATM inside premises

These represent two types of ATM installations: on and off premise. On premise ATMs are typically more advanced, multi-function machines that complement an actual bank branch's capabilities and thus more expensive. Off premise machines are deployed by financial institutions and also ISOs (Independent Sales Organizations) where there is usually just a straight need for cash, so they typically are the cheaper mono-function devices. In Canada, when an ATM is not operated by a financial institution it is known as a "White Label ATM". In North America, banks often have drive-thru lanes providing access to ATMs. Many ATMs have a sign above them indicating the name of the bank or organization owning the ATM, and possibly including the list of ATM networks to which that machine is connected. This type of sign is called a topper.

Most ATMs are connected to interbank networks, enabling people to withdraw and deposit money from machines not belonging to the bank where they have their account or in the country where their accounts are held (enabling cash withdrawals in local currency). Some examples of interbank networks include PULSE, PLUS, Cirrus, Interac, Interswitch, STAR, and LINK.

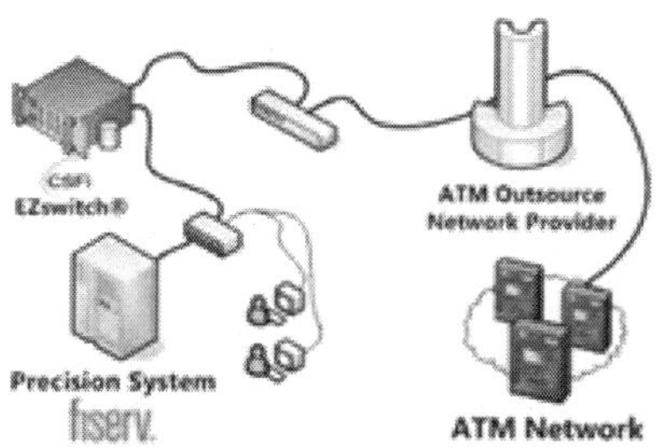

ATM connections

ATMs rely on authorization of a financial transaction by the card issuer or other authorizing institution via the communications network. This is often performed through an ISO 8583 messaging system.

Many banks charge ATM usage fees. In some cases, these fees are charged solely to users who are not customers of the bank where the ATM is installed; in other cases, they apply to all users. Where machines make a charge some people will not use them, but go to a system without fees.

In order to allow a more diverse range of devices to attach to their networks, some interbank networks have passed rules expanding the definition of an ATM to be a terminal that either has the vault within its footprint or utilizes the vault or cash drawer within the merchant establishment, which allows for the use of a scrip cash dispenser.

ATMs typically connect directly to their ATM Controller via either a dial-up modem over a telephone line or directly via a leased line. Leased lines are preferable to POTS lines because they require less time to establish a connection. Leased lines may be comparatively expensive to operate versus a POTS line, meaning less-trafficked machines will usually rely on a dial-up modem. That dilemma may be solved as high-speed Internet VPN connections become more ubiquitous. Common lower-level layer communication protocols used by ATMs to communicate back to the bank include SNA over SDLC, TC500 over Async, X.25, and TCP/IP over Ethernet.

In addition to methods employed for transaction security and secrecy, all communications traffic between the ATM and the Transaction Processor may also be encrypted via methods such as SSL.

Although ATMs were originally developed as just cash dispensers, they have evolved to include many other bank-related functions. In some countries, especially those which benefit from a fully integrated cross-bank ATM network. ATMs include many functions which are not directly related to the management of one's own bank account, such as:

- Deposit currency recognition, acceptance, and recycling
- Paying routine bills, fees, and taxes (utilities, phone bills, social security, legal fees, taxes, etc.)
- Printing bank statements
- Updating passbooks
- Loading monetary value into stored value cards
- Purchasing
- Postage stamps.
- Lottery tickets
- Train tickets
- Concert tickets
- Movie tickets
- Shopping mall gift certificates.
- Games and promotional features

- Donating to charities
- Cheque Processing Module
- Adding pre-paid cell phone credit.

1.2 Reliability

Before an ATM is placed in a public place, it typically has undergone extensive testing with both test money and the backend computer systems that allow it to perform transactions. Banking customers also have come to expect high reliability in their ATMs, which provides incentives to ATM providers to minimize machine and network failures. Financial consequences of incorrect machine operation also provide high degrees of incentive to minimize malfunctions.

ATMs and the supporting electronic financial networks are generally very reliable, with industry benchmarks typically producing 98.25% customer availability for ATMs and up to 99.999% availability for host systems. If ATMs do go out of service, customers could be left without the ability to make transactions until the beginning of their bank's next time of opening hours.

This said, not all errors are to the detriment of customers; there have been cases of machines giving out money without debiting the account, or giving out higher value notes as a result of incorrect denomination of banknote being loaded in the money cassettes. Errors that can occur may be mechanical (such as card transport mechanisms; keypads; hard disk failures); software (such as operating system; device driver; application); communications; or purely down to operator error. To aid in reliability, some ATMs print each transaction to a roll paper journal that is stored inside the ATM, which allows both the users of the ATMs and the related financial institutions to settle things based on the records in the journal in case there is a dispute. In some cases, transactions are posted to an electronic journal to remove the cost of supplying journal paper to the ATM and for more convenient searching of data.

Improper money checking can cause the possibility of a customer receiving counterfeit banknotes from an ATM. While bank personnel are generally trained better at spotting and removing counterfeit cash, the resulting ATM money supplies used by banks provide no absolute guarantee for proper banknotes, as the Federal Criminal Police Office of Germany has confirmed that there are regularly incidents of false banknotes having been dispensed through bank ATMs. Some ATMs may be stocked and wholly owned by outside companies, which can further complicate this problem. Bill validation technology can be used by ATM providers to help ensure the authenticity of the cash before it is stocked in an ATM; ATMs that have cash recycling capabilities include this capability.

1.3 Fraud

As with any device containing objects of value, ATMs and the systems they depend on to function are the targets of fraud. Fraud against ATMs and people's attempts to use them takes several forms.

The first known instance of a fake ATM was installed at a shopping mall in Manchester, Connecticut in 1993. By modifying the inner workings of a Fujitsu model 7020 ATM, a criminal gang known as The Bucklands Boys were able to steal information from cards inserted into the machine by customers.

In some cases, bank fraud could occur at ATMs whereby the bank accidentally stocks the ATM with bills in the wrong denomination, therefore giving the customer more money than should be dispensed. The result of receiving too much money may be influenced on the card holder agreement in place between the customer and the bank.

In a variation of this, WAVY-TV reported an incident in Virginia Beach of September 2006 where a hacker who had probably obtained a factory-default admin password for a gas station's white label ATM caused the unit to assume it was loaded with $5 USD bills instead of $20s, enabling himself—and many subsequent customers—to walk away with four times the money they said they wanted to withdraw. This type of scam was featured on the TV series The Real Hustle.

ATM behavior can change during what is called "stand-in" time, where the bank's cash dispensing network is unable to access databases that contain account information (possibly for database maintenance). In order to give customers access to cash, customers may be allowed to withdraw cash up to a certain amount that may be less than their usual daily withdrawal limit, but may still exceed the amount of available money in their account, which could result in fraud.

1.3.1 Card fraud

In an attempt to prevent criminals from shoulder surfing the customer's PINs, some banks draw privacy areas on the floor.

Fig Hacking of ATM card

For a low-tech form of fraud, the easiest is to simply steal a customer's card. A later variant of this approach is to trap the card inside of the ATM's card reader with a device often referred to as a Lebanese loop. When the customer gets frustrated by not getting the card back and walks away from the machine, the criminal is able to remove the card and withdraw cash from the customer's account.

Another simple form of fraud involves attempting to get the customer's bank to issue a new card and stealing it from their mail.

The concept and various methods of copying the contents of an ATM card's magnetic stripe on to a duplicate card to access other people's financial information was well known in the hacking communities by late 1990.

In 1996 Andrew Stone, a computer security consultant from Hampshire in the UK, was convicted of stealing more than £1 million (at the time equivalent to US$1.6 million) by pointing high definition video cameras at ATMs from a considerable distance, and by recording the card numbers, expiry dates, etc. from the embossed detail on the ATM cards along with video footage of the PINs being entered. After getting all the information from the videotapes, he was able to produce clone cards which not only allowed him to withdraw the full daily limit for each account, but also allowed him to sidestep withdrawal limits by using multiple copied cards. In court, it was shown that he could withdraw as much as £10,000 per hour by using this method. Stone was sentenced to five years and six months in prison.

By contrast, a newer high-tech modus operandi involves the installation of a magnetic card reader over the real ATM's card slot and the use of a wireless surveillance camera or a modified digital camera to observe the user's PIN. Card data is then cloned onto a second card and the criminal attempts a standard cash withdrawal. The availability of low-cost commodity wireless cameras and card readers has made it a relatively simple form of fraud, with comparatively low risk to the fraudsters.

In an attempt to stop these practices, countermeasures against card cloning have been developed by the banking industry, in particular by the use of smart cards which cannot easily be copied or spoofed by un-authenticated devices, and by attempting to make the outside of their ATMs tamper evident. Older chip-card security systems include the French Carte Bleue, Visa Cash, Mondex, Blue from American Express and EMV '96 or EMV 3.11. The most actively developed form of smart card security in the industry today is known as EMV 2000 or EMV 4.x.

EMV is widely used in the UK (Chip and PIN) and other parts of Europe, but when it is not available in a specific area, ATMs must fallback to using the easy to copy magnetic stripe to perform transactions. This fallback behaviour can be exploited. However the fallback option has been removed by several UK banks, meaning if the chip is not read, the transaction will be declined.

In February 2009, a group of criminals used counterfeit ATM cards to steal $9 million from 130 ATMs in 49 cities around the world all within a time period of 30 minutes.

Card cloning and skimming can be detected by the implementation of magnetic card reader heads and firmware that can read a signature embedded in all magnetic stripes during the card production process. This signature known as a "MagnePrint" or "BluPrint" can be used in conjunction with common two factor authentication schemes utilized in ATM, debit/retail point-of-sale and prepaid card applications.

An ATM allows a bank customer to conduct their banking transactions from almost every other ATM machine in the world. DON Wetzel was the co-patentee and chief conceptualist of the ATM. The concept of the ATM first began in 1968, a working prototype came about in 1969 and Docutel was issued a patent in 1973. The first working ATM was installed in a New York based Chemical Bank.

The World's first ATM was installed in Enfield Town in the London Borough of Enfield, London on June 27-1 967 by Barclays Banks. This instance of the invention is credited to john Sheoherd-Barron although George Simjian registered patents in New York, USA in the 1930.

The First Voucher Based cash dispensing machine was installed in 1967 by Barclay's Bank in London. And then the next version was going to be the total teller created in 1971, which is the ATM we all know today.

1.3.2. Invention

It is true that most inventions have happened due to sheer necessity and invention of ATM machine is one of them. Now the question that arises is what was the necessity, what called for the invention of a round the clock cash dispenser? Who invented it- a broke student or a shopaholic lady, a businessman or a banker, too tired of cashing the cheques? Also known as Cashpoint or Hole-in-the-Wall Machine (Britain), ABM or Automatic Banking Machine (USA), All-time Money (India), and Minibank (Norway) and so much more, the history of ATM is full of interesting facts, some we know, some we don't.

An Armenian named Luther George Simjian was forced to move to USA in the year 1920, under the account of Armenian Genocide. He owned to his credit the invention of a portrait camera and then rolled out the formulated idea of ATM, the Automated Teller Machine.

Confident of his invention, he persuaded Citibank to run his product on a six month trial basis. Soon enough, he was disappointed with the performance and the lack of users and concluded that ATM was a wasteful addition to personal banking. And lack of demand for the ATM finally forced him to take a back seat. Clear enough; the time was not right for this concept to have been accepted generously. Simjian clearly lost out on the success and fame and the same was passed on to two other gentlemen, John Shepherd-Barron and Don Wetzel.

John Shepherd-Barron was a Scottish national born in India. Later he relocated to Britain and pursued his education from the University of Edinburgh, and at Trinity

College, Cambridge. After returning empty handed from the bank, Shepherd-Barron was disappointed to have no other solution to wait till the bank would open next. And thus in a similar fashion like Archimedes, Shepherd-Barron claims to have hit his Eureka moment while taking a bath. A self-sufficient cash dispensing machine was what he was thinking about. And soon the ATM was invented in the early 1960s. The invention of a self-sufficient cash dispensing machine was his second and successful attempt at inventions. Earlier he had invented an instrument to scare away seals at his Scottish Salmon farms. Unfortunately, this device instead of deterring the seals attracted them, and was a failure.

The ATM machine gained Shepherd-Barron an ever-lasting recognition in the banking world and paved the way for hi-tech banking techniques, online bank accounts and PIN and chip security technology. The four-digit internationally accepted standard PIN was also invented by him. Earlier, he had a six-digit Army serial number in his mind but later his wife suggested for a shorter PIN as it would be easy to remember. Finally in 1967 that the first ATM that dispensed paper currency round the clock, was unveiled. The ATM machine installed outside a Barclay's bank in North London started dispensing cash on a 24 hour basis.

As the plastic cards were still to have come into existence, this machine accepted and generated money through cheques impregnated with certain chemicals. Majorly a mild radioactive substance, Carbon 14 was used for detection by the machine. Once the PIN was given, the machine gave out the cash. This radioactive substance had no ill effects on the health of users and Shepherd-Barron claimed that a user would have to eat about 136,000 cheques to suffer any kind of ill-effects. Reg Varney, a famous TV sitcom popular became the first person to use the ATM in the year 1967 and withdrew about 10 dollars. The amount seems too less for us, but this money was enough for a complete night out spent on the tiles in London, inclusive of dinner, drinks, a show and a taxi-ride back to home, in short enough cash for a "Wild Weekend".

While this prototype device originated by Shepherd-Barron had started functioning, various parallel developments were happening in different parts of the world. An American Engineer Donald Wetzel of Docutel engineered the Docuteller

ATM which was declared as the first modern magstripe machine. It recognized magnetically encoded plastic (credit cards) and not the usual paper cheques.

And there have been a lot of efforts gone into final development of the ATM, the ones we see today, the ones we use so frequently, and the ones which have made our lives revolve around plastic money. The development of ATM ever since its baby steps in the late 1930s and then gearing up for longer runs in the 1960s, and finally a matured and stable stage that we see the ATMs in today. Undoubtedly, most of the ideas and patents contributed for makeover of the ATM from time to time form the backbone of what was initiated as "holes in the wall".

Today, ATMs hold a strong foothold in the world, offering everyone a better access to their money, be it in any corner of the world. Let's put figures to assumptions, there are about 1.8 million ATMs in use around the world with ATMs on cruise and navy ships, airports, newsagents and petrol stations. ATMs too have been categorized as on and off premise ATMs. On Premise ATMs are capable to connect the users to the bank with multi-function capabilities. Off premise, ATM machines on the other hand are the "white label ATMs" and are limited to cash dispense, no balance enquiries, no statement print-out.

The developments have not stopped; the contactless technology is on its rise. Shepherd-Barron continued to take inimitable and lively interest in technology well even in his old age and had foreseen a future where plastic cards too would be numbered. For his excellent and unforgettable contributions to financial technologies, he was also offered the OBE in the year 2005. And in the year 2010, he took his last breath and left behind his legacy of technological advancements which would refuses to end. Many more inventions are in process and many will be successful too. The time is just right to bring in the glorious inventions rolling in.

1.3.3 Inventors

Luther George Simjian

In the late 1930's, Luther George Simjian started building an earlier and not-so-successful version of an ATM. He did register related patents. He initially came up with the idea of creating a "hole-in-the-wall machine." It would allow customers to make financial transactions, without entering the bank. The idea was met with a great deal of doubt. Starting in 1939, Simjian registered 20 patents related to the device and persuaded what is now Citicorp to give it a trial. After six months, the bank reported that there was little demand.

John Shepherd-Barron

John Shepherd-Barron had an idea in the 1960's for a 24/7 cash dispenser. He was managing director of De La Rue Instruments. De La Rue today manufactures cash dispensers. There is a De La Rue cash dispenser in 1 out of every 5 ATM machines built. If you want to say that Shepherd-Barron invented the ATM, then the world's first ATM was installed outside North London. It was installed in a branch of Barclays Bank in 1967. Later that year, Shepherd-Barron presented his idea to a conference. The conference consisted of 2,000 US bankers in Miami, this was after the first ATMs had been installed in England. He spoke to the conference about the new self-service banking device he developed. On December 31, 2004, John Shepherd-Barron, was named an OBE, Officer of the Order of the British Empire by the Queen. It was for "services to banking." Shepherd Barron said: "It was a bit late, but better late than never." Press releases stated that Shepherd-Barron was the "Inventor of the ATM." But, was he really the inventor?

James Goodfellow

As a Development engineer with Smiths Industries Ltd, James Goodfellow was given a project. It was to develop an automatic cash dispenser in 1965. Chubb Lock & Safe Co. were to provide the secure physical housing. They were also to develop a mechanical dispenser mechanism. Mr. Goodfellow designed the new system. It consisted of a machine readable encrypted card, and the machine that read it. To this he added a numerical keypad. UK Patent No.1,197,183 with a priority date of May 2 1966, covers this invention. It is also covered by US Patent No.3,905,461 and Patents granted by many other countries. These Patents list James Goodfellow as inventor, along with the late A.I.O.Davies, the company General Manager. This US Patent still describes the basic ATM function almost 40 years later. These Machines were marketed by Chubb LTD. They were installed nationwide in the UK during the late 60s and early 70s.

Don Wetzel

In 1968, according to a NMAH interview, Don Wetzel, says he was the Vice President of Product Planning at Docutel. The company that developed automated baggage-handling equipment. He applied for a patent on an ATM machine. He said there were two other inventors listed on the patent. They were Tom Barnes, a mechanical engineer. George Chastain, an electrical engineer. It took five million dollars to develop their ATM according to Mr. Wetzel.

John D. White

John D. White told ATMmachine.com that his work started in 1968. He told us that he installed the first ATM at Rockville Center, LI for the then Chemical Bank in August 1973. His design was patented on May 9, 1973 for the Docutel Corporation and was filed on July 29, 1970. The machine was a "Credit Card Automatic Currency Dispenser". Mr. White provided copies of his patent to ATMmachine.com. We reviewed them. It does indeed state that the inventor of the machine was John D. White and Kenneth Goldstein. The assignee on the patent was the Docutel Corporation. This is very convincing evidence. It appears that it was White and not Wetzel who received the patent. There is also a statement in the patent that supports the idea of the modern ATM.

"Both the original code and the updated code are scrambled in accordance with a changing key", which is what happens today. ATM's are programmed with security keys and the code changes. Today they must be 3DES certified. It is encrypted to prevent fraudulent access to credit card and ATM debit numbers. This occurs between the machine, the bank, and the network processor.

Jairus Larson

Jairus Larson told ATMmachine.com he did not invent the ATM. But, as far as he is aware, he developed the first 'on-line' ATM. This would be Diebold's "550". The first ATM's were all 'off-line' versions (referred to in the past as 'stand-alone'). Meaning they did not have any means to communicate with the bank network. Today's ATMs are "on-line". On-line in the sense that they communicate with the bank's computer system.

1.4. History

1.4.1 USA 1968-1969

An automatic teller machine or ATM allows a bank customer to conduct their banking transactions from almost every other ATM machine in the world. Don Wetzel was the co-patentee and chief conceptualist of the automated teller machine, an idea he thought of while waiting in line at a Dallas bank. At the time (1968) Wetzel was the Vice President of Product Planning at Docutel, the company that developed automated baggage-handling equipment. The other two inventors listed on the patent were Tom Barnes, the chief mechanical engineer and George Chastain, the electrical engineer. It took five million dollars to develop the ATM. The concept of the ATM first began in 1968, a working prototype came about in 1969 and Docutel was issued a patent in 1973. The first working ATM was installed in a New York based Chemical Bank. There are different claims to which bank had the first ATM, here is Don Wetzel's reference:

The first voucher based cash dispensing machine was installed in 1967 by Barclay's Bank in London. Experts, however, do not consider this an ATM. The first modern day ATM was introduced to consumers in 1969 by Chemical Bank.

"No, it wasn't in a lobby, it was actually in the wall of the bank, out on the street. They put a canopy over it to protect it from the rain and the weather of all sorts. Unfortunately they put the canopy too high and the rain came under it. (laughing) One time we had water in the machine and we had to do some extensive repairs. It was a walkup on the outside of the bank. That was the first one. And it was a cash dispenser only, not a full ATM... We had a cash dispenser, and then the next version was going to be the total teller (created in 1971), which is the ATM we all know today -- takes deposits, transfers money from checking to savings, savings to checking, cash advances to your credit card, takes payments; things like that. So they didn't want just a cash dispenser alone." - Don Wetzel on the first ATM installed at the Rockville Center, New York Chemical Bank from a NMAH interview.

Fig: Early model of ATM

Jack Gebhart had the idea to put the magnetic stripe on the bankcard to carry the customer information. This information could then be read by a reader, also from Jack Gerbhart, which in its turn is connected to a computer.

1.4.2 Timeline

1939

Bank automation started as early as 1939 with the invention of the ATM by Luther George Simjian (1905-1997) from Turkish origin.

Simjian came up with the idea of creating a hole-in-the-wall machine that would allow customers to make financial transactions. The idea met skepticism but he persuaded what is now Citicorp to give it a trial. After six months, the bank reported that there was little demand. "It seems the only people using the machines were a small number of prostitutes and gamblers who didn't want to deal with tellers face to face," wrote Simjian.

1960

ATM predecessor installed: New York's First National City Bank (now CitiBank) installs a Bank graph in several branch lobbies. The idea is for customers to pay utility bills and get receipts without having to see a teller.

1967

- Cash dispensing machine at Barclay's Bank in London on June 27. On June 27 1967 De La Rue wins the race to unveil the first machine, drawing crowds to the unveiling.

- The concept is developed by John Shepherd- Barron. He came up with a self-service machine which dispenses paper currency on a 24/7 basis. The first of this type of machines will be installed outside a north London branch of Barclays Bank in 1967.
- It uses paper vouchers bought from tellers. The machine is called the De La Rue Automatic Cash System, or DACS.

1968

- The concept of the ATM established by Wenzel.
- Card-eating: Barclays and a few other banks introduce a machine that encodes cash on plastic cards purchased from a teller. The problem is that the machine always eats the cards, and customers have to buy new cards if they want to make more transactions.

1969

- First use of ATM magstripe cards: Docutel installs its Docuteller machine at New York's Chemical Bank. The installation marks the first use of magnetically encoded plastic.

Fig Docuteller machine

- Chemical Bank's ad campaign announces: "On Sept. 2, our bank will open at 9:00 and never close again!"
- Other manufacturers get into the game, but Docutel is the first to apply for a patent. Docutel is later credited by the Smithsonian's National Museum of American History as being the ATM's inventor. But most people in the industry recognize Docutel's first machine as the first modern magstripe machine.
- Donald C. Wetzel is given credit for developing the Docutel machine, which is initially met with resistance from bankers. Bankers say they're concerned about the machine's cost, which is about $8,000 more annually than a human teller. The bankers also fear that customers won't like having a machine handling their money.

1971

- First true bank ATMs: Docutel introduces its Total Teller, the first true fully functioning bank ATM.
- Omron (Japan) installs the first On-line cash dispenser.

Fig On-line cash dispenser

1973

- Docutel rewarded a patent for the ATM.
- Proliferation begins: By 1973, 2,000 ATMs — mostly from Docutel- are operating in the United States. They sell for about $30,000 each.
- Diebold begins work on the development of its TABS 600 with plans for an off-line prototype.

Fig Diebold TABS 500

1974

- Diebold's first TABS 500 machine, an off-line version, is installed at a bank in Atlanta.
- Diebold develops the TABS 550, an on-line and off-line version. The flexibility of the machine makes it unique and more marketable to banks.
- The TABS 600 also is developed.

1977

Chungho ComNet Co., Ltd. begins developing Teller Machines for banks.

1978

The first IBM-compatible Diebold machine is installed at a bank in Indianapolis. (8)

1979

Chungho ComNet Co (Korea) begins selling Cash Dispensers

1983

- ❖ Chungho ComNet Co (Korea) begins selling On-line Teller Machines. Machines like this now have a direct connection via telephone lines with the bank's mainframe computers.
- ❖ In later years this connection is regarded as too high risk open for fraud and the machines' records will only be updated once to several times a day.

1994

- ❖ ICL, the IT services company, has designed the first cash dispenser (CD) with a bill paying facility.
- ❖ The new CD incorporates a barcode scanner, which allows customers to pay bills automatically from their bank accounts, as well for traditional cash withdrawals and to check account balances.
- ❖ Banque du Caire, Egypt's third largest bank, is the first organisation to adopt the new technology in a bid to reduce queues in branches and offer customers a more convenient service. ICL will design and integrate 100 new cash dispensers for the bank in a deal worth £2.4million.

The first people to be able to use the service will be customers of an Egyptian national mobile phone network. Their phone bills will incorporate a barcode holding details of the amount due, as well as their bank account number at the Banque du Caire. The customer simply sweeps the barcode over the scanner at the cashpoint, keys in their PIN number and the bill is paid.

1.5. Evolution

ATMs are such a regular part of our everyday lives; that are available in shopping malls and in workplaces.

1960-1967: Technological run-ups to ATM machine

- ❖ The first ATM machine predecessors were invented out of a need for after-hours banking.
- ❖ Workers were frustrated that they couldn't pay bills or send money because banks were closed by the time they knocked off work.
- ❖ The first to actually look (but not function) like a present-day ATM machine

was the Bankograph (installed in 1961 in New York by City Bank) that didn't dispense cash but accepted it – allowing the public to pay their bills and send money without having to interact with a cashier or teller.

- The Bankograph wasn't very popular and was off the market only six months after its release. But the effects of its introduction to the public would linger for years afterwards.
- Patents were filed by various companies between 1962 and 1967 for machines that would be able to dispense money to bank account holders without them having to go inside the bank.

1967: The first cash dispenser is installed

- On 27 June 1967, in Enfield Town, London, the first cash dispensing machine was installed by Barclays Bank known as De La Rue Automatic Cash System or DACS (named for De La Rue, the printing firm its inventor worked for).
- Users insert cheques into the machines and the machines would return the appropriate amount.
- The cheques were called tokens and they were treated with an isotope of carbon that the machine could read and interpret securely.
- The tokens were mailed back to users after the transactions were processed.
- Though not ideal by today's security standards, this system was quite popular at the time and would eventually inspire the use of plastic bank cards in ATMs.

1969-1972: The first modern ATMs are deployed

- In 1969, Chemical Bank deployed the first ATM machine to actually look and work like modern ATMs; the Docuteller, manufactured by the Docutel Corporation, premiered in New York on 2 September.
- Docutel were the first to hold the USA patent for the ATM and are therefore generally credited as its inventor.
- A string of machines were released by several other companies that marked the proliferation of ATMs into the market which are widely considered as first true ATMs, because they were the first to dispense cash using bank-issued cards that worked in combination with security keys (which we now know as PIN numbers).

1979 – late1990s: Innovation takes over

- After ATMs were installed by major banks all over the world, all that was left to do was make them better.
- When the internet came into wide use, around the late 90s, the next step was to connect ATMs to the internet so that they could update automatically and quickly.

- Diebold and NCR were the only two manufacturers competing the networked ATM space.
- Over a time, many other manufacturers, like Triton, who haven't stopped innovating and have recently released their Argo ATM range with features like anti-skimming technology, resistive and capacitive touch screens with e-receipts joined.
- From new functionality and first ATMs that used wireless technology rather than fixed line, to experimentations with dye stain technology to mitigate theft of cash, ATMs haven't stopped evolving since they were invented.
- ATMs are now being dropped anywhere, as they can operate off solar power and inform merchants of low cash volumes.
- The next frontier for ATM machine technology appears to be biometrics, as fingerprint-scanning ATMs hurtle closer and closer to being a market standard.

1999: ATM Solutions is born

- Eighteen years after the first ATM went live in South Africa, ATM Solutions was founded.
- At the time, the problem was simple: ATMs could only be found outside bank branches, so were often too far away from where people lived, shopped and worked which coincides with rapid change in retail landscape.
- With this, ATM Solutions embarked on 17-year journey of business and innovation focused on bringing people closer to their money.
- Previously, ATM access was often inconvenient, but Paycorp's ATMs meant that a retailer of any size could have an ATM installed – inside or outside – bringing a new level of convenience to users.
- Paycorp installed 5 000 ATMs in South Africa, Namibia, Zambia with a fledgling ATM business in Hungary.

Chapter 2
Features of Automated Teller Machine (ATM)

2.1 Introduction

Automated Teller Machine is also referred to as a cash machine, a cash dispenser and 'the hole in the wall' among other names. The ATM is an electronic computerized telecommunications device that allows financial institutions (e.g. bank or building society) customers to directly use a secure method of communication to access their bank accounts. The ATM is a self-service banking terminal that accepts deposits and dispenses cash. Most ATM's also let users carry out other banking transactions (e.g. check balance). ATM's are activated by inserting a bank card (cash or credit card) into the card reader slot. The card will contain the customers account number and PIN (Personal Identification Number) on the cards magnetic stripe. When a customer is trying to withdraw cash for example, the ATM calls up the banks computers to verify the balance, dispenses the cash and then transmits a completed transaction notice.

Fig ATM cash machine

The idea for an ATM originally was to simply replace or reduce the workload of a bank teller (i.e. the person in the bank who gives out money to customers). The ATM would help reduce banks overheads as wages would be decreased. As for who created the first ATM or where it was first used is a topic of much debate. Basically what answer you get when the question 'who invented the ATM?' is asked depends on who you ask. Miller in 2006 presents the facts as he knows it about the history and invention of the ATM. The notion of having a bank machine which automatically dispensed cash to customers came about in the 1930's.

A Turkish born inventor working in America called George Simijan started building an earlier and not-so-successful version of an ATM in the late 1930's. He registered the related patents. Simijan came up with the idea of a 'hole-in-the-wall' machine which would allow customers to make financial transactions. However, at the time this idea was well ahead of its time and was met with great doubt. Simijan registered 20 patents related to the device and persuaded an American bank to trial it. However, after 6 months the bank reported little demand in the service and it was withdrawn. It was not until the 1960's that the idea of the ATM was looked at again. John Shepherd-Barron, an inventor from the UK, had an idea in the 1960's for a 24/7 cash dispenser. At the time Shepherd-Barron was the managing director of a company called De La Rue Instruments which today still manufactures cash dispensers. People who believe John Shepherd-Barron invented the ATM argue that the worlds first ATM was installed outside a north London branch of Barclays in 1967.

In 1965 a Scottish man called James Goodfellow was given a project to develop an automatic cash dispenser. Goodfellow was a development engineer with a UK company called Smiths Industries Ltd. He designed a system which accepted a machine readable encrypted card and had a numerical keypad used to enter a PIN. This design is covered in patents in both the UK and USA among other countries. This patent still describes the basic ATM function 40 years later (i.e. the design was patented in 1966). Goodfellow's machines were marketed by Chubb Ltd and installed throughout the UK during the late 1960's and early 1970's. Don Wetzel, then the Vice President of Product Planning of the American Corporation Docutel, claims he applied for a patent on an ATM in 1968. In fact some people believe Wetzel to be the inventor of the ATM. However, an ATM design patented in 1973, stating the Docutel Corporation as the assignee, states John D White as the inventor. White claims he started working on ATM system in 1968 and he installed the first ATM in 1973. This machine was called the 'Credit Card Automatic Currency Dispenser'.

Fig Credit Card Automatic Currency Dispenser

Evidence suggests it was White who received the patent and not Wetzel. There is also a statement in the patent which supports the idea of the modern ATM – "Both the original code and the updated code are scrambled in accordance with a changing key". This is basically what happens today. ATM's have security keys programmed into them. The code changes and is scrambled to prevent access to credit and ATM card numbers between the ATM, the bank and the network processor. It is clear that the topic of ATM invention is quite a controversial one. However, the combined effort of all the inventors surely has helped create today's ATM. Anyone who worked on ATM design from the 1930's until today has contributed something to the modern ATM designs. The purpose of this research is to investigate existing ATM design and to design a 'best of breed' ATM user interface design.

2.1.1 Reliability

Before an ATM is placed in a public place, it typically has undergone extensive testing with both test money and the backend computer systems that allow it to perform transactions. Banking customers also have come to expect high reliability in their ATMs, which provides incentives to ATM providers to minimize machine and network failures. Financial consequences of incorrect machine operation also provide high degrees of incentive to minimize malfunctions.

ATMs and the supporting electronic financial networks are generally very reliable, with industry benchmarks typically producing 98.25% customer availability for ATMs and up to 99.999% availability for host systems. If ATMs do go out of service, customers could be left without the ability to make transactions until the beginning of their bank's next time of opening hours.

Of course, not all errors are to the detriment of customers; there have been cases of machines giving out money without debiting the account, or giving out higher value notes as a result of incorrect denomination of banknote being loaded in the money cassettes. Errors that can occur may be mechanical (such as card transport mechanisms; keypads; hard disk failures); software (such as operating system; device driver; application); communications; or purely down to operator error.

To aid in reliability, some ATMs print each transaction to a roll paper journal that is stored inside the ATM, which allows both the users of the ATMs and the related financial institutions to settle things based on the records in the journal in case there is a dispute. In some cases, transactions are posted to an electronic journal to remove the cost of supplying journal paper to the ATM and for more convenient searching of data.

Improper money checking can cause the possibility of a customer receiving counterfeit banknotes from an ATM. While Bank presonnel are generally trained better at spotting and removing counterfit cash, the resulting ATM money supplies used by banks provide no absolute guarantee for proper banknotes, as the Federal

Criminal Police Office of Germany has confirmed that there are regularly incidents of false banknotes having been provided through bank ATMs. Some ATMs may be stocked and wholly owned by outside companies, which can further complicate this problem when it happens. Bill validation technology can be used by ATM providers to help ensure the authenticity of the cash before it is stocked in an ATM; ATMs that have cash recycling capabilities include this capability.

2.2 Interacting with ATMs

To access the funds which are kept in the bank at any time is not an easy task but today it is not at all difficult. A person just has to tell his bank that he wants an ATM card. The bank issues him an ATM card which is passcoded and could be used by him alone. ATM card is called by different names like bank card, MAC(Money access card), client card, key card or cash card, etc. In most cases even debit and credit card could be used.

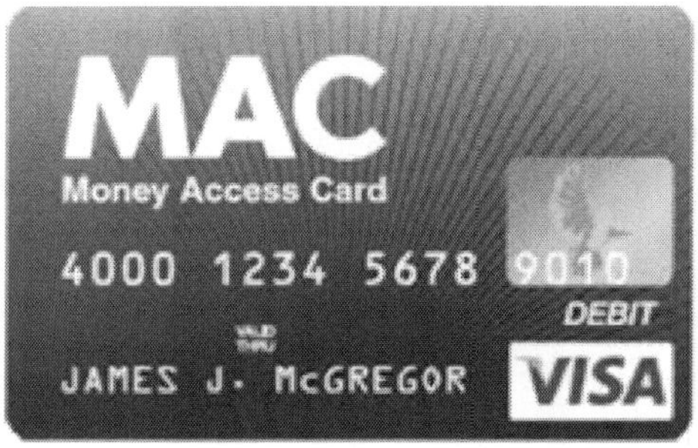

Fig Money access card

The ATM card helps the customer to be identified by a plastic ATM card with a magnetic stripe or a plastic smartcard with a chip. The security is provided by the customer entering a personal identification number (PIN). Now the person who needs funds when he is taking his family out of shopping or when he requires funds for an outing it is easily done. If someone falls very seriously ill the person can remove the funds in the middle of the night too.

Although ATM's provide an extremely useful service to banks customers, at times they can be very frustrating to use and therefore there is a lot of room for

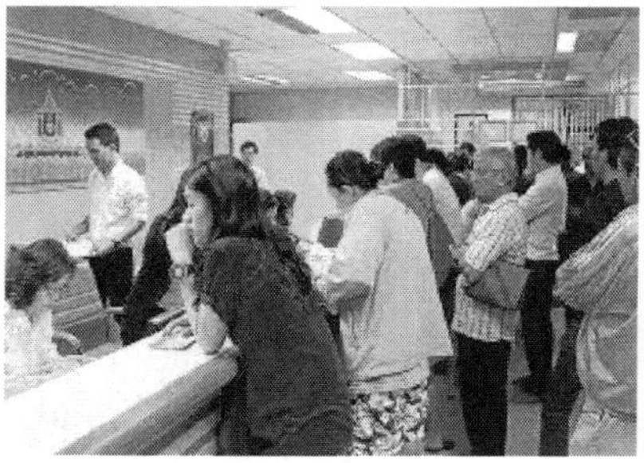

Customer withdrawing cash from Bank and ATM

improvement in the interface design. The interface enables communication between the user and the machine. Therefore good user interface design is imperative for high usability levels. Often there are problems or inconveniences experienced when using an ATM. Some of these problems include:

- Waiting in the queue to use the ATM. If users ahead of you in the queue experience difficulties in using the machine, this will increase the time waiting in the queue.
- Inability to see the ATM screen well. This depends on the location of the ATM in relation to the position of the sun. At times it can be difficult to view the contents of the ATM menu.
- Wrongly inserting the ATM card. This problem is more common with new ATM users who are not familiar with their new card and the ATM.
- Getting the required amount of money. Some ATM's may not offer the user the required amount of money they want on the initial cash withdrawal screen. The user will then have to use a few more key strokes to select the required amount (e.g. to withdraw £50 the user might have to select the 'other amounts' option then type in '50' using the keypad and then press 'enter').
- Understanding how to perform operations. Some ATM users find the instructions on how to perform operations quite difficult to understand.
- Often the ATM card is returned to the user while further operations are required (e.g. the card is returned once the user requests a sum of cash. However the user may want to do further transactions; such as check balance or top-up a mobile phone). This will lead to the customer having to re-insert their ATM card, further increasing their time spent at the ATM.

On some ATM machines the menu options are not aligned with their corresponding menu key. An example of this is illustrated in Figure.

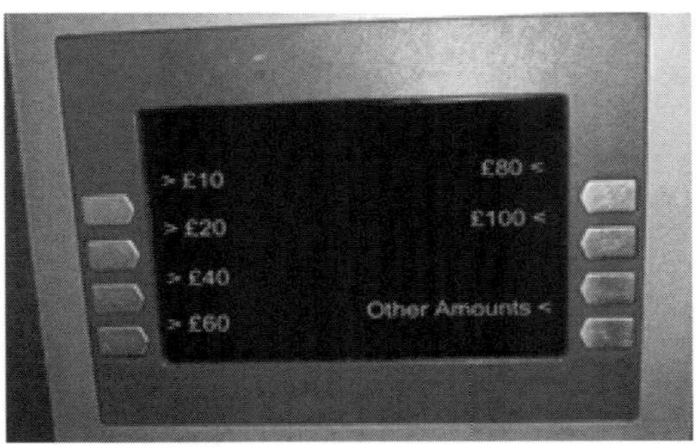

Fig: Typical ATM Menu

Although the sums of money £10 to £100 are not aligned with the related keys, most users will be able to determine what keys are to be used to select the required sum of money. However, if a user wanted to select the 'Other Amounts' option; what button is to be pressed? There is obvious reason for confusion here. It is evident that problems exist with the use of ATM's. Some of these problems are unavoidable (e.g. an ATM running out of money) but solutions exist for others. This research paper focuses on the user interface design problems. ATM navigation menus could be improved considerably to make ATM's more usable.

As technology increases the ATM interface should evolve to take advantage of the new technological innovations. This has happened to a certain extent over the years. However, it is clear that most of today's ATM interfaces do not have the desired high level of usability they should. The modern ATM should be flexible, expressive and easier to use. As mentioned earlier ATM's were introduced in the UK in the late 60's and early 70's. ATM's can now be found in shops, hotels and airports among other places. There was a major design problem when ATM's were first introduced.

During a transaction the ATM dispensed cash to the customer before returning the customers card. This resulted in customers not collecting their card from the ATM. This design problem has now been rectified. The customers' card is returned before cash is dispensed. There have been improvements in the usability of ATM's over the years but there is still a lot of room for improvement. The modern ATM is much more than a simple cash dispenser. Standard UK ATM's offer relatively basic services including cash withdrawals; balance checks and the ability to top-up pay-as-you-go mobile phones. ATM's in different countries tend to offer advanced services which include cash deposits, cheque deposits, paying bills, purchasing tickets (e.g. train, concert) and purchasing stamps.

The design of an ATM should not only include its inherent usability but also its perceived usability'. This is just one version of possible problems encountered when using (or trying to use an ATM). It reinforces the problems that exist with ATM use. Another typical problem, which was already mentioned earlier, is when an ATM returns the customers card prematurely i.e. the user still has additional transactions to make. This process is as:

- Insert card
- Enter PIN
- Choose transaction option (Withdraw cash)
- Select/Enter amount of cash to be withdrawn
- Receipt? (yes/no)
- Card ejected from ATM
- Take cash
- Re-insert card
- Enter PIN

- ❖ Choose transaction option (Balance Enquiry)
- ❖ Return card

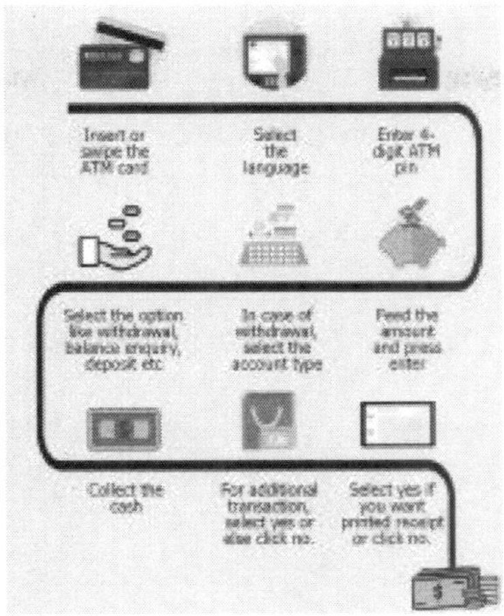

ATM steps

This shows how using an ATM can be frustrating. Human computer interface is a term used to describe the interaction between a user and a computer; in other words, the method by which a user tells the computer what to do, and the responses which the computer makes. Also states Human-Computer Interaction (HCI) is about designing computer systems that support people so that they can carry out their activities productively and safely. This can be summarised as 'to develop or improve the safety, utility, effectiveness, efficiency and usability of systems that include computers'. If ATM's were more usable then they would become more effective and efficient machines as users would find them easier to use. This would cause the users to spend less time using the machines and to carry out more efficient transactions. This would be very desirable as it would lessen waiting times in a queue to use an ATM's services. This research paper is concerned with the usability of ATM's; to investigate why existing ATM's user interfaces (navigation menus in particular) have problems and to design a proposed 'best of breed' ATM menu system with excellent usability.

Preece explains usability is concerned with making systems easy to learn and easy to use. Poorly designed computer systems can be extremely annoying to users. This point is particularly relevant. ATM's, at times, can be extremely annoying to use for many reasons which were mentioned earlier. In order to produce computer systems with good usability HCI specialists strive to understand the factors that determine how people operate and make use of the computer technology effectively; develop tools and techniques to help designers ensure that computers systems are suitable for the activities for which people will use them and achieve efficient, effective and safe interaction both in terms of individual human-computer interaction and group interactions.

The last point is relevant for ATM design as users want their banking interactions to be as quick as possible. However, using an ATM's services is very personal (especially with the development of ATM crime) so the group interactions can be ignored in this case. A good interface design can help to ensure that users carry out task when the using the system:

❖ Safely – this is important for safety-critical software systems; such as software for a jumbo jet for example.

❖ Effectively – the user get what they want from the system e.g. if an ATM user requests £100 cash, the user should get this and not £50.

❖ Efficiently – this is the main point concerned with this research paper. If the ATM menu's were improved this would make ATM use more efficient. For example users don't want to spend 5 minutes trying to find the correct way to insert their cash card and type their PIN and the amount of cash they want and then eventually leave without remembering to extract their cash card.

❖ Enjoyably – systems should be attractive and inviting. Generally if a system is effective and efficient to use, it should also be enjoyable to use as a consequence. However additional effort could be made in ATM interface design to make ATM's more enjoyable to use such as making the screens and menus more colourful and have images for example. A lot of ATM's still just have a black background screen with illuminated text, which is quite dull.

Well designed systems can improve systems significantly. They can improve the output of employees, improve the quality of life and make the world a safer and enjoyable place. An ATM is a service a bank offers to its customers. There are two factors which contribute to the usage of a particular ATM. These are location and the usability of the ATM. Obviously location is the major factor. If an ATM is conveniently located then it will be used a lot. If an ATM is easy to use then this will encourage customers to use the ATM. Many people may have preferences over other ATM's and if they had the choice would use their preferred ATM all the time. All in all, the greater usage a banks ATM receives, the more potential there is for the bank to make profit. This is why a bank or building society should not under-estimate the importance of good ATM interface design.

Preece states that 'the best user interface design guidelines are guidelines in a true sense: high level and widely applicable directly principles'. The following principles can be applied widely:

❖ Know the user – This can often be difficult to achieve, especially when a diverse population of users has to be accommodated or when the users can only be anticipated in the most general terms. This is particularly true for ATM user interface design as this system has a wide range of users from teenagers to pensioners.

❖ Reduce Cognitive Load – This concerns designing so that users do not have to remember large amounts of detail. Again this is very relevant for ATM user interface deign. The ATM system should be easy to use and users should remember how to understand how to use the system.

- Engineer for errors – a system should be designed to accommodate inevitable user error. If the user makes an error while using the system the system should be able to recover. Engineering for errors includes taking forcing actions to try and prevent users from making errors initially, providing good error messages, and using reversible actions to apply users to correct their own errors.
- Maintain consistency and clarity – Consistency emerges from standard operations and representations and from using appropriate metaphors that help to build and maintain a user's mental model of a system. For example the 'desktop' in a PC is an appropriate metaphor of a work desktop in an office. ATM user's interfaces generally use consistent language e.g. withdraw cash, PIN services etc. However, different banks offer different ATM user interfaces. It would be ideal if there was a universal ATM user interface design, or at least a standard design in each country.

A number of studies have already been carried out regarding ATM's. Most of these studies however have focused on ATM use in relation to the age of users and user disabilities (such as blindness). Adams and Thieben, Mead et al., Rogers et al. and Rogers and Fisk concentrate on ATM use in relation to the age group of the users. Mankze et al. focuses on ATM usability by the blind while Hone et al. focuses on modes of control for ATM's including voice control. Rogers et al. say that they have been informed by banking staff that training is not necessary for ATM's because they are inherently user friendly. This statement however is often not true as many people find ATM's difficult to use, never mind the elderly users and users who are disabled in some way (for example blind). There has also being significant research done on ATM usability and user behaviour. Each ATM investigated (one from each bank e.g. Bank of Ireland, First Trust etc) is evaluated and measured by efficiency (transaction times). This is done using 'mock-up' ATM prototypes which are direct replicas of the Bank's ATM menu designs.

2.3 ATM Applications

There are many ATM applications which are described as:

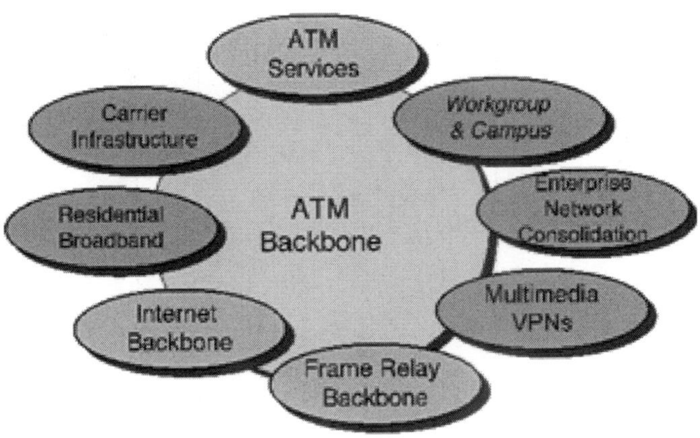

ATM services—Service providers globally are introducing or already offering ATM services to their business users.

ATM workgroup and campus networks—Enterprise users are deploying ATM campus networks based on the ATM LANE standards. Workgroup ATM is more of a niche market with the wide acceptance of switched-Ethernet desktop technologies.

ATM enterprise network consolidation—A new class of product has evolved as an ATM multimedia network-consolidation vehicle. It is called an ATM enterprise network switch (ENS). A full-featured ATM ENS offers a broad range of in-building (e.g., voice, video, LAN, and ATM) and wide-area interfaces (e.g., leased line, circuit switched, frame relay, and ATM at narrowband and broadband speeds) and supports ATM switching, voice networking, frame-relay SVCs, and integrated multiprotocol routing.

Multimedia virtual private networks and managed services—Service providers are building on their ATM networks to offer a broad range of services. Examples include managed ATM, LAN, voice and video services (these being provided on a per-application basis, typically including customer-located equipment and offered on an end-to-end basis), and full-service virtual private-networking capabilities (these including integrated multimedia access and network management).

Frame-relay backbones—Frame-relay service providers are deploying ATM backbones to meet the rapid growth of their frame-relay services to use as a networking infrastructure for a range of data services and to enable frame relay to ATM service interworking services.

Internet backbones—Internet service providers are likewise deploying ATM backbones to meet the rapid growth of their frame-relay services, to use as a networking infrastructure for a range of data services, and to enable Internet class-of-service offerings and virtual private intranet services.

Residential broadband networks—ATM is the networking infrastructure of choice for carriers establishing residential broadband services, driven by the need for highly scalable solutions.

Carrier infrastructures for the telephone and private-line networks—Some carriers have identified opportunities to make more-effective use of their SONET/ SDH fiber infrastructures by building an ATM infrastructure to carry their telephony and private-line traffic.

2.4 ATM Accessing Points

- Hotels and Motels.
- Convenience Stores.
- Shopping Centers.

- Restaurants, Bars and Pubs.
- Banks.
- Commercial Areas.
- Residential localities.
- Major Petrol Pumps.
- Airports
- Near Railway Stations, etc.

2.5 Features of ATM

ATM Features

- Cash withdrawal: Normally allowing up to Rs.25000/- per day from Customers' account (50,000/- for HNI's)
- Balance Enquiry: Know our ledger balance and available balance.
- Mini Statement: Get a printout of our last 8-10 transaction on our Current Account.
- Deposit Cash/Cheque: Avails at all full function ATMs customer can deposit both cash and cheques. Cash deposited in ATMs will be credited to the account on the same day (providing cash in deposited before the clearing) and cheques are sent for clearing on the next working day.
- Fund Transfer: Transfers funds form one account to another linked account in the same branch and in different bank account.
- PIN Change: Change the Personal Identification Number of ATM.
- Payments: The latest feature of ATMs is used for payment of bills, making donations to temples/ trusts, buying internet packs, airtime recharges for prepaid mobile phones, and much more.
- Others: Request for a cheque book from ATMs and our concerned branch will dispatch in such that it reaches you within 10 working days.

2.6 ATM Services

1. Mobile Prepaid Card Recharge

   Use your ATM Card at any of the Bank availing this service: we can purchase mobile re-charge cards.

2. Internet Packs (ATN - Any Time Net)

   We can purchase new internet connection or buy or renewal packs by bank ATMs.

3. Donations

All Bank Debit / ATM Cardholders can now refill their prepaid phone cards through their bank ATMs or through SMS. This facility is available for all circles of the following operators:

a. Hutch/ Orange

b. Airtel

c. Idea.

4. Charity

Through this facility we can donate for charity against any of the following heads:

a. Child Welfare.

b. Education.

c. Health & Sanitation.

d. Livelihoods & Poverty.

e. Service for the disabled.

5. Calling Cards

Calling cards enable a user to make a call either through the normal phone or through the Internet protocol. There are two kinds of cards available.

a. India on All Cards

b. Hello World Cards

2.7 ATM Cards

The Cardholder or the customer can access cash, make payments at Merchant Establishments, ascertain information about his Account Balance through the use of the Cards at ATMs devices (whether of the Bank or Shared Network) for viewing Account information through Corporate Interest Banking/for using Corporate Care

Services or any such services as specified by the Bank from time to time. The card is valid for use at ATMs and Merchant Establishments in India and abroad. However, the Card is not valid for payment in foreign exchange at Merchant Establishments in India. The Card is valid up to the last working day of the month indicated on the Card.

The advantage of almost universal acceptability of a debit card comes with a subtle disadvantage of indulging in extravagance without thought of prudence. The ATM card, while ensuring that you do not spend beyond limits also comes with the inconvenience of hunting for an ATM Center every time you need cash.

When someone opens a deposit account with a bank, he/she will frequently be given a payment card to access the funds in the account. In most cases, the card will be a debit card or check card, or, in the case of a line of credit, a credit card. In addition to making purchases, all of these cards function as ATM cards, meaning that they can be used to withdraw funds at an ATM.

Some banks may issue an ATM-only card, which can't be used to make purchases; it can only be used to perform regular ATM functions, although some exceptions apply, such as when the payment processing company has a terminal installed at the retailer. All debit cards are ATM cards, but not all ATM cards are debit cards.

As with other payment cards, ATM cards are secured with a personal identification number (PIN). ATMs allow customers to deposit and withdraw cash as well as check their balance, but customers may have to pay a fee when using an ATM card at an ATM not associated with the issuing financial institution.

In today's advanced era, ATM card has become a need, as it is the fastest source to carry out any relevant financial transaction. It can be used for several purposes, hence helps top keep records and consume time. As of the advantages, the card also has its own disadvantages; hence there are methods which can stop the card from any misuse.

ATM is a payment card provided by a financial institution to their customers; it enables the customer to use an automated teller machine (ATM) for transactions

Fig Debit Card

such as: deposits, cash withdrawals, obtaining account information, and other types of banking transactions, often through inter-bank networks. Some ATM cards are also used for paying loans and getting cash advances, however ATM cards cannot be used for making purchases. There are several different types of cards that fall under the category of an ATM card.

An ATM debit card can function similar to that of a normal ATM card, with an additional functionality of debiting or removing the cash from any account. ATM debit cards can be used for purchasing purposes in online mode also. They look similar to that of ATM credit cards, but does not work like them. These cards are used for removing money purposes.

Fig ATM Credit Cards

This type of card has all features of both, the ATM debit card and that of the credit card. It's one among the new types of debit cards available in some banks. Along with the checking account, the card is also attached to the line of credit. Here, if the person overdrafts their accounts, then they will not obtain any of the fees back; instead the money from their account would be taken off with interest when the funds would be available.

Some banks also offer ATM and ATM debit cards, which are used in the form of Picture ID. This is done for the security purposes, a small picture can be added on the card after a relevant payment. Besides these added features, there are also options to change the appearance of the card. Hence, even the ATM cards can be designed according to the customer's wish or desire.

Besides these there are also, combo cards available in the financial institutes, which functions as per their policies and rules. ATM card have gained importance in recent year, due to its conformability and speed. The technology has made easy to access the money from any ATM centre, enabling an easy accesses.

2.7.1 Features

The benefits of using Bank ATM cards are more than evident.

Fig ATM card

Save time

This is perhaps the single most benefit. Save time in driving down to the branch and you do not have to waste time waiting in the queue to perform your transactions. The ATM Card facilities provide you with the option of banking instantly for various transactions mentioned above.

Convenient 24×7 Banking

At the ATM, now you are not bound to do your transactions within banking hours. There no need to worry about bank holidays or public holidays. It is 24 x 7, 365 days a year banking facility.

Withdraw Cash when overseas

Depending on the type of facility provided by your bank, if you are travelling overseas, then the card can be used to draw currency of the country you are travelling to, from the ATM. Of course, you should check before hand if your bank has given you this facility.

Universally Accepted

Whether you are in town or out of town (or in another country), you can forget about carrying loads of cash with you. Just carry your ATM cum debit card. It is normally accepted everywhere. However you must check with the restaurant or the hotel if they have a system of accepting the credit and debit cards. You can almost be sure if there is mastercard logo on your debit card. Also you should make double sure by telling your bank that you're leaving your station and there should be any interruption in the service.

Security

The use of ATM is restricted only to the person who knows the PIN, so if you keep the PIN confidential then no one but you can use the ATM to do the banking transactions. Also for safety purposes, you can set daily transaction limits and the ATM receipts can help you keep track of your transactions and the money withdrawn or transferred. Also you can change your PIN as an additional security measure.

Save your account from being dormant

Often one or the other account goes dormant if you have not done any transaction for 6 months or so. To make it operational again is a hassle. The reason for this is that most of us are hard pressed for time. Now you can do one small transaction at regular intervals to keep it active.

Helpful Budgeting Tool

With an ATM Debit card you can never go overboard with your money spending. The debit card ensures that you get to spend only that much money that you have in your bank account. This means you cannot go into a situation like "credit card debt". This helps you to strictly stay within the budget limits, unlike with credit cards.

2.8 Consumer Benefits

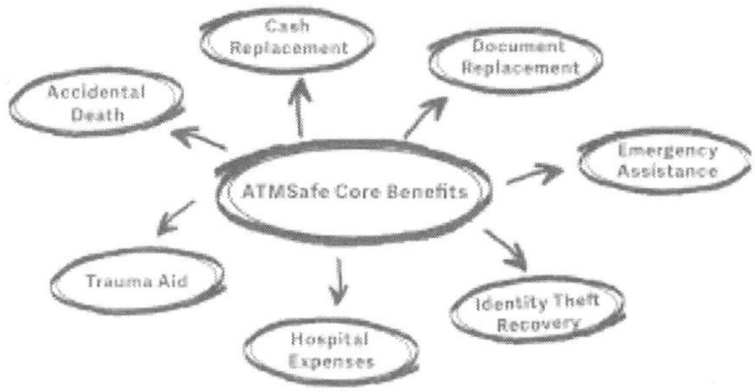

- ❖ All the convenience of a bank card for persons currently having no banking relationships.
- ❖ Accepted at gas stations, supermarkets, department stores, restaurants and places of entertainment.
- ❖ Benefits of not carrying cash with the convenience associated with bank account
- ❖ Security of PIN.
- ❖ Balance recovery by cardholder in case of lost card.
- ❖ Convenience, easy to use and easier to carry than cash.
- ❖ Immediate anytime access to account information.

2.9 Advantages

- ❖ ATM provides 24 hours service.
- ❖ ATM gives convenience to the bank's customers.

- ATM reduces the workload of bank's staff.
- ATM provides service without any error.
- ATM is very beneficial for travellers as they don't have to carry large amount of cash.
- ATM may give the consumers new currency notes and not soiled notes.
- It reduces the hustle and bustle which is involved in a transaction when done through teller.
- ATM provides privacy while conducting banking transactions.
- The bank customers expect high reliability in their ATM's which provide incentive to the bank to minimise machine and network failures.
- There is no need of filling forms because ATM requires no such document.
- The customer is able to access the bank from any part of the world and can conduct essential banking services like deposits, withdrawals, transfer of funds, etc.
- A lot of expansion of services can happen with an ATM to any corner of the world by providing electronic access to its customers.
- ATM's help in reducing the cost of operation as it reduces human intervention and increases profitability of banks.
- The financial consequences which would arise of an incorrect machine also provides incentive to the bank to minimise malfunctions.
- Some ATM's print each transaction to a roll paper journal which is stored in the ATM. This allows the users and the related banking institution to settle things based on records if any dispute arises.
- For customer security the lobbies have extensive security camera coverage, a courtesy telephone for consulting with the bank staff and a security guard on the premises.
- Openings on the customer-side of ATMs are often covered by mechanical shutters to prevent tampering with the mechanisms when they are not in use.
- Alarm sensors are placed inside the ATM and in ATM servicing areas to alert their operators when doors have been opened by unauthorised person.

2.10 Disadvantages

- ATM Card cloning fraud
- Fake ATM's
- Attack on ATM's
- PIN crack

- If ATM networks do go out of service, customers could be left without the ability to make transactions until the beginning of their bank's next time of opening hours.
- The ATM machine doesn't guarantee a 100% availability of cash. It may run low of funds and the customers have to wait till the management takes care of it.
- The cost levied to a customer using an ATM can be higher.
- Robbers preyed on people using money machines in poorly lit or otherwise unsafe locations and criminals also devised ways to steal customers' PINs.
- Consumers were faced with an increase in ATM crimes and scams.
- ATM's cannot be provided in rural areas: In a country like India, where banks are having large number of rural and non-computerized branches, ATM services become difficult to be provided.
- The presence of various constraints makes it more difficult to introduce ATM services in the country side like illiteracy, security concern, etc.
- There is a limitation of cash withdrawals from ATM's. Many banks do not permit withdrawal of more than 25,000 at a time.
- The cash deposit facility is restricted and not safe as dropping of envelope in an ATM is not advisable.
- There is strong possibility of misusing ATM card if misplaced, lost or stolen. There are number of such reported incidences nowadays.
- There is loss of personnel touch of customers with the Banks.

2.11 Responsible Functions of ATM

Planning infrastructures and services

- ATM plans collective public transport infrastructures and programmes those that must be carried out within a timescale of ten years and, at the same time, defines their characteristics, programmes the necessary investment, supervises the corresponding projects and draws up the planning instruments to coordinate the Metropolitan Collective Public Transport System, including the Intermodal Transport Plan, if necessary.
- ATM draws up investment programmes for periods of more than one year and proposes agreements to finance infrastructures that must be signed by the state central government (AGE) and the Catalan government, as well as the instruments required to cover the completion of other investments carried out via other formulas.
- It also monitors the execution of current investments in the system contained in the programming instruments, whether such investments come from the infrastructure financing agreements taken out with the AGE and the Generalitat de Catalunya or through other financing channels.

- It plans the services and establishes coordinated programmes for their use by all the companies providing these services and monitors trends in the global mobility market with special attention to trends in private transport.

Relations with collective transport operators

- Via delegation or jointly with service owners, ATM takes out programme contracts or other kinds of agreements with the public transport service providers in its geographical area.
- It draws up the programme contracts that must be taken out between ATM and operators.
- It monitors the agreements and management contracts of all the private and public firms of the
- Metropolitan Collective Public Transport System.
- It monitors developments and compliance of programme contracts in force with Transports Metropolitans de Barcelona (Barcelona Metropolitan Transport Authority) and with Ferrocarrils de la Generalitat de Catalunya (Catalan Railway Authority).
- It draws up and proposes agreements with Renfe (Spanish Railway Authority) regarding the integration of the local railways services provided by Renfe in Barcelona within Metropolitan Collective Public Transport System.

Financing the system via public administrations

- ATM draws up proposals for financing agreements with the different public administrations responsible for financing public transport and defines the budgetary allocations to the system.
- It takes out financing agreements with public administrations to offset the deficit of services and operational costs of the management structure.
- It controls income, costs and investments by service providers for several purposes.

Organising tariffs

- It has joint responsibility for the tariffs charged by administrations owning collective transport and, among these, is the system's sole liaison with the economic administration.
- It draws up and approves a shared system of tariffs as part of a financing policy that defines to what extent revenue must cover the costs and establishes the integrated tariff system.
- It owns the revenue from tariffs for combined tickets and redistributes the amount required among the operators.

Communication

- ATM disseminates the corporate image of the Metropolitan Collective Public Transport System and of ATM itself, with respect and in accordance with the images of the service owners and operators.

- ❖ It carries out communication campaigns in order to encourage the use of the aforementioned system among the population.
- ❖ It's responsible for advertising, information and user relations.

Future regulatory framework

- ❖ ATM studies and draws up proposals for the administrations in the consortium (individually or via the groups or bodies that represent them before ATM) regarding the adaptation of the regulatory framework to help the consortium exercise the powers established as part of its area of authority.
- ❖ It exercises administrative powers regarding the organisation of the services within its geographical area, powers which are attributed to ATM by the public administrators signing the framework agreement of 28 July 19995 and other affiliated organisations.
- ❖ It establishes relations with other administrations to encourage due compliance of the functions attributed to ATM in accordance with its statutes.

Other mobility-related functions

- ❖ ATM draws up, processes and evaluates master plans for mobility.
- ❖ It issues reports on urban mobility plans, service plans and studies evaluating general mobility.
- ❖ It applies and finances measures to encourage the rational use of private vehicles, notwithstanding government powers at a local and autonomous community level.
- ❖ It draws up and finances proposals to encourage the rational use of public highways and spaces related to aspects such as parking, pedestrian areas and the implementation of lanes reserved for public transport or bicycles, notwithstanding government powers at a local and autonomous community level.
- ❖ It fosters a culture of sustainable mobility among the public.
- ❖ It carries out any other function attributed to it by current legislation in the area of mobility.

2.12 Technological developments

Corporate information systems are generally constructed to support business activities. In contrast, the systems used by banks are directly linked to the products and services offered by the banks. This means the quality of the infrastructure established by each bank has a major effect on how the bank itself is regarded. Financial institutions headed the transition to second-generation online systems that began in Japan in the mid-1970s.

In 1972, planning its second-generation online system ahead of other banks, Fuji Bank asked OKI to develop a new terminal system. The bank wanted to renovate its existing system, which handled only deposits and currency exchange, into a more sophisticated online system capable of processing the full range of bank operations.

In response, OKI developed the OKITAC-1300 terminal system in 1973. While the first-generation online system required the host computer to perform all processes for terminal units, the OKITAC-1300 used terminal controllers (TCs) installed between the host computer and terminal units. Equipped with memory (65 Kbytes) as capacious as found in the mid-sized computers of the time, the TC performed input/output data check and editing. Each TC could control the operations of several branch offices, reducing the workload on the host computer. The OKITAC-1300 also allowed the connection of more terminals, contributing to widespread use of online systems by financial institutions.

In November 1975, Nippon Cash Service ("NCS"), a company established jointly by city banks, local banks, and mutual banks, began operating cash dispensers (CDs). NCS installed CDs in train stations, department stores, and other locations, spurring rapid diffusion of CDs. Immediately after ATMs appeared in 1979, the market for automated teller machines expanded sharply.

Fig First CD delivered to Fuji Bank

OKI delivered the first CD machine to Fuji Bank in December 1970. In 1974, its cumulative sales of CDs reached 2,000 units. That year, OKI began developing automatic depositors (ADs). In this effort, the know-how accumulated during the development of CDs proved useless. Since the banknotes customers pulled from their billfolds varied significantly in condition, ADs had to process severely worn banknotes that human bank tellers would not ordinarily hand to customers. These machines had particular difficulty separating banknotes. To handle this latter task, OKI developed a new rotary-type suction mechanism incorporating a vacuum pump. OKI faced another difficult challenge: To keep a second banknote from sticking to the first banknote. After repeated trial and error, OKI developed a mechanism incorporating a rubber belt rotating in the reverse direction to separate the second and subsequent banknotes.

Financial institutions installed many ATMs in their cash service corners during the 1980s. These ATMs tended to feature two cassettes: one for deposited banknotes and another for banknotes to be dispensed. Each ATM contained a large amount of cash. OKI began examining efficient systems that would allow deposited banknotes to be immediately used for customer withdrawals.

Fig ATM manufactured by OKI

The most difficult part of this development was to neatly place all the deposited banknotes, ranging so widely in condition, in the stacker (cashbox) for storage, and then to dispense these banknotes from the stacker one by one for withdrawals. The mechanism required to handle this task (and thereby develop the world's first cash-recycling ATM) took OKI roughly one year to develop. In 1982, OKI unveiled the AT-100 series ATMs, featuring an all-new banknote acceptance and dispensing system. The AT-100 series included models offering various novel functions, including bank transfer using a color display and envelope acceptance functions. Automated bank transfer machines added functions that allowed them to accept coins and return change; this allowed customers to use both cash and bank cards to transfer money.

Today, ATMs can be found all over Japan, including most convenience stores. This state of affairs is based on the many obstacles overcome during the early years of ATM development.

2.13 ATM Working

The automated teller machine is simply a data terminal with two input and four output devices. These devices are interfaced to the processor. The processor is heart of the ATM machine. All the ATM machines working around the world are based on centralized database system.

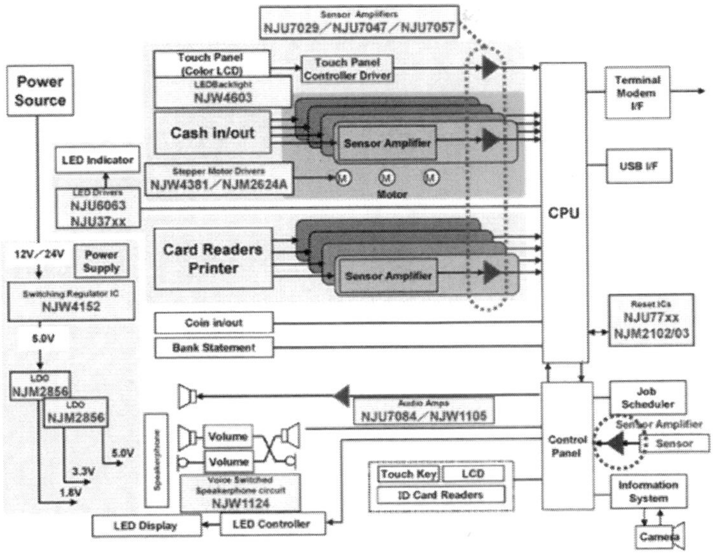

Fig Automatic Teller Machine Process

The ATM has to connect and communicate with the host processor (server). The host processor is communicating with the internet service provider (ISP). It is the gateway through all the ATM networks available to the card holder.

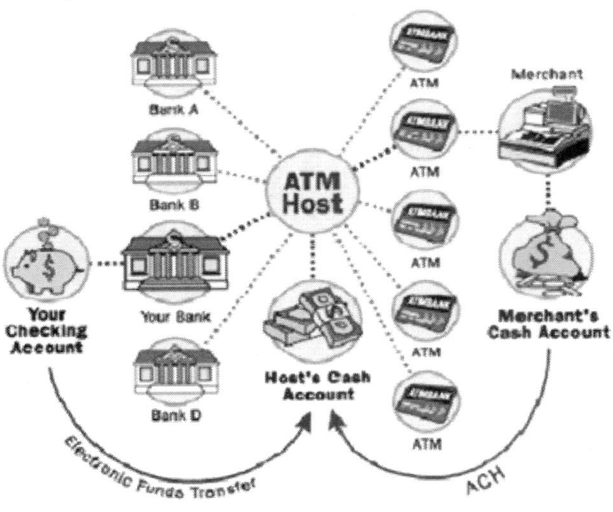

Fig Automatic Teller Machine Architecture

When a card holder wants does an ATM transaction, user provides necessary information through card reader and keypad. The ATM forwards this information to the host processor. The host processor enters the transaction request to the cardholder bank. If the card holder requests the cash, the host processor takes the cash from the card holder account. Once the funds are transferred from the customer account to host processor bank account, the processor sends approval code to the ATM and the authorized machine to dispense the cash. This is the way to get the amount on ATMs. The ATM network is fully based on centralized database environment. This will make life easier and secured the cash.

2.13.1 ATM Process

State-Chart for One Session

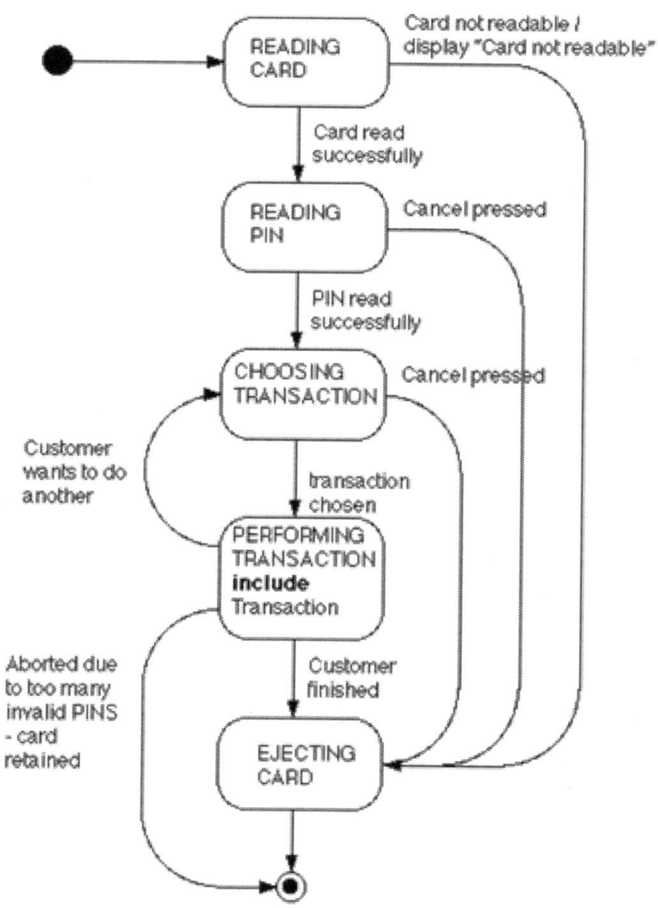

Chapter 3

Design and Installation of Automated Teller Machine (ATM)

3.1 Introduction

The automated teller machine (ATM) is an automatic banking machine (ABM) which allows customer to complete basic transactions without any help of bank representatives. There are two types of automated teller machine (ATMs). The basic one allows the customer to only draw cash and receive a report of the account balance. Another one is a more complex machine which accepts the deposit, provides credit card payment facilities and reports account information.

It is an electronic device which is used by only bank customers to process account transactions. The users access their account through special type of plastic card that is encoded with user information on a magnetic strip. The strip contains an identification code that is transmitted to the bank's central computer by modem. The users insert the card into ATMs to access the account and process their account transactions. The automated teller machine was invented by john shepherd-Barron in year of 1960.

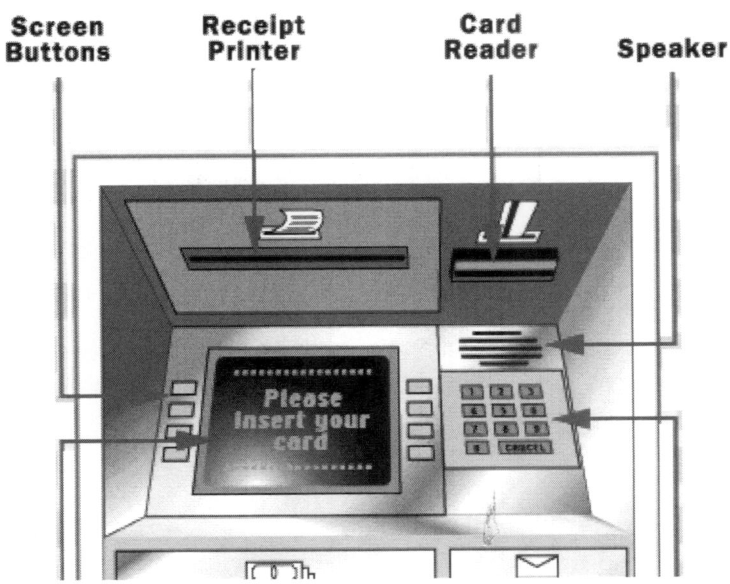

Fig ATM

3.2 Block Diagram

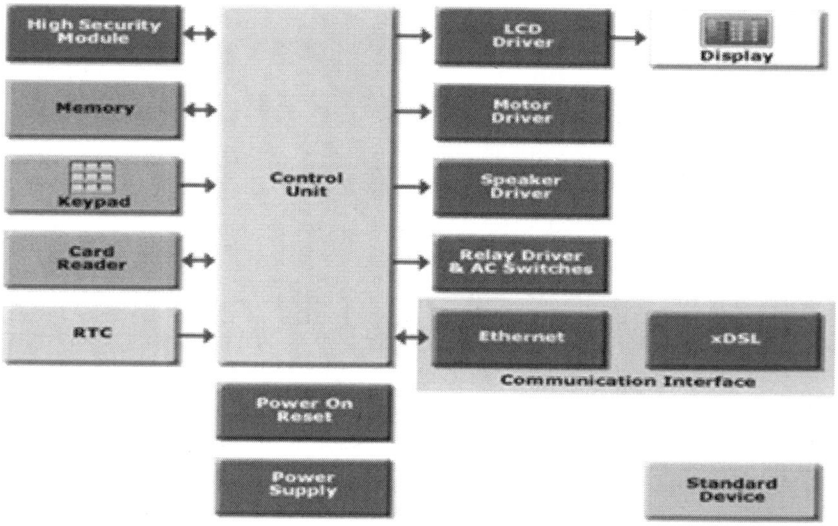

Fig Block diagram of ATM

The automated teller machine externally consists of mainly two input devices and four output devices that are;

3.2.1 Input Devices:

- Card reader

The card reader is an input device that reads data from a card .The card reader is part of the identification of your particular account number and the magnetic strip on the back side of the ATM card is used for connection with the card reader. The card is swiped or pressed on the card reader which captures your account information i.e. the data from the card is passed on the host processor (server). The host processor thus uses this data to get the information from the card holders.

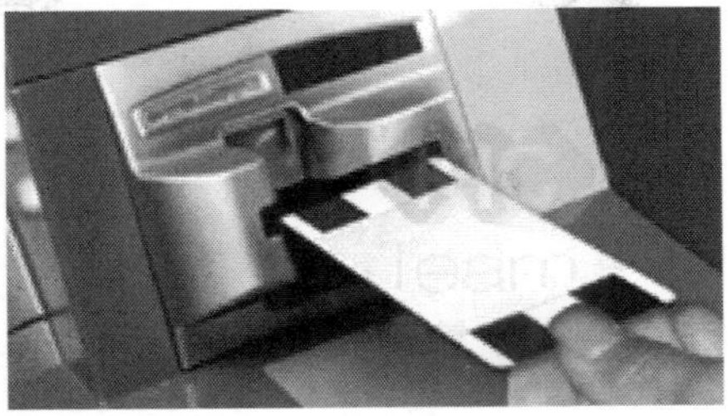

Fig Card Reader

Responsibilities

- ❖ Tell ATM when card is inserted
- ❖ Read information from card
- ❖ Eject card
- ❖ Retain card
- ❖ Keypad

The card is recognized after the machine asks further details like your personal identification number, withdrawal and your balance enquiry Each card has a unique PIN number so that there is little chance for some else to withdraw money from your account. There are separate laws to protect the PIN code while sending it to host processor. The PIN number is mostly sent in encrypted from. The key board contains 48 keys and is interfaced to the processor.

Fig Keypad

3.2.2 Output Devices:

- ❖ Speaker

The speaker provides the audio feedback when the particular key is pressed.

- ❖ Display Screen

The display screen displays the transaction information. Each steps of withdrawal is shown by the display screen. A CRT screen or LCD screen is used by most of ATMs.

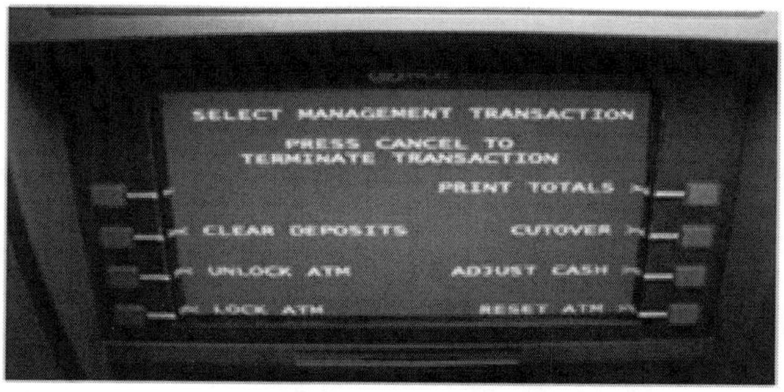

Fig Display Screen

- **Receipt Printer**

The receipt printer print all the details recording your withdrawal, date and time and the amount of withdrawn and also shows balance of your account in the receipt.

- **Cash Depositor**

The cash dispenser is a heart of the ATM. This is a central system of the ATM machine from where the required money is obtained. From this portion the user can collect the money. The duty of the cash dispenser is to count each bill and give the required amount. If in some cases the money is folded, it will be moved another section and becomes the reject bit. All these actions are carried out by high precision sensors. A complete record of each transaction is kept by the ATM machine with help of an RTC device.

Fig ATM cash dispenser

Responsibilities

- Keep track of cash on hand, starting with initial amount
- Report whether enough cash is available
- Dispense cash

3.3 Hardware Parts in ATM

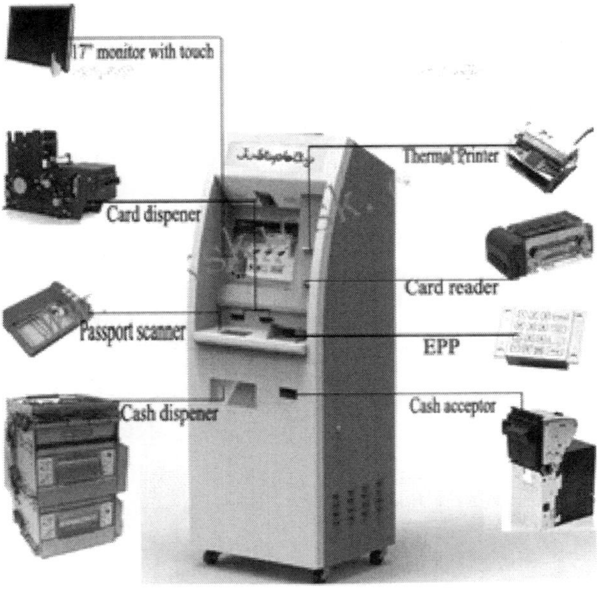

An ATM is typically made up of following devices:

- ❖ CPU (to control the user interface and transaction devices)
- ❖ Magnetic and/or Chip card reader (to identify the customer)
- ❖ PIN Pad (similar in layout to a Touch tone or Calculator keypad), often manufactured as part of a secure enclosure.
- ❖ Secure cryptoprocessor, generally within a secure enclosure.
- ❖ Display (used by the customer for performing the transaction)
- ❖ Function key buttons (usually close to the display) or a Touchscreen (used to select the various aspects of the transaction)
- ❖ Record Printer (to provide the customer with a record of their transaction)
- ❖ Vault (to store the parts of the machinery requiring restricted access)
- ❖ Housing (for aesthetics and to attach signage to)

Recently, due to heavier computing demands and the falling price of computer-like architectures, ATMs have moved away from custom hardware architectures using microcontrollers and/or application-specific integrated circuits to adopting a hardware architecture that is very similar to a personal computer. Many ATMs are now able to use operating systems such as Microsoft Windows and Linux. Although it is undoubtedly cheaper to use commercial off-the-shelf hardware, it does make ATMs vulnerable to the same sort of problems exhibited by conventional computers.

Business owners often lease ATM terminals from ATM service providers. The vault of an ATM is within the footprint of the device itself and is where items of value are kept. Scrip cash dispensers do not incorporate a vault. Mechanisms found inside the vault may include:

- Dispensing mechanism (to provide cash or other items of value)
- Deposit mechanism, including a Cheque Processing Module and Batch Note Acceptor (to allow the customer to make deposits)
- Security sensors (Magnetic, Thermal, Seismic)
- Locks: (to ensure controlled access to the contents of the vault)
- Journaling systems; some are electronic (a sealed flash memory device based on proprietary standards) or a solid-state device (an actual printer) which accrues all records of activity, including access timestamps, number of bills dispensed, etc. - This is considered sensitive data and is secured in similar fashion to the cash as it is a similar liability.

ATM vault

ATM vaults are supplied by manufacturers in several grades. Factors influencing vault grade selection include cost, weight, regulatory requirements, ATM type, operator risk avoidance practices, and internal volume requirements.

Industry standard vault configurations include Underwriters Laboratories UL-291 "Business Hours" and Level 1 Safes, RAL TL-30 derivatives, and CEN EN 1143-1:2005 - CEN III/VdS and CEN IV/LGAI/VdS. ATM manufacturers recommend that vaults be attached to the floor to prevent theft.

3.3.1 Virtual Channel Identifier / Virtual Path Identifier

ATM provides two types of transport connection, Virtual paths and Virtual channels. A virtual channel is a unidirectional pipe made up from the concatenation of a sequence of connection elements. A virtual path consists of a set of these channels

- The virtual circuit or VC (Virtual Channel) is a generic term used to describe the communication capacity for the transport of ATM cells. A virtual circuit identifier, VCI or, conventionally called commutation reference, is assigned to a VC connecting transporting ATM cells between two ATM nodes. The ATM node, wherein the VCI value is translated, is also called an ATM switch.
- The virtual circuit connection, or VCC (Virtual Channel Connection), defines the end to end connection between two access points to the AAL. a VCC is composed by concatenating one or more VC.
- The VP or VP (Virtual Path) is a VC beam. All VC of a bundle have the same end nodes.
- The virtual path connection, or VPC (Virtual Path Connection), consists of the concatenation of one or more VP. The ATM node is then called a fan.

Figure shows the VP / VC hierarchy and Figure Examples of VPC and VCC. The journey of a VPC is established by a routing in intermediate brewers (A, B and C for VP1 in Figure).

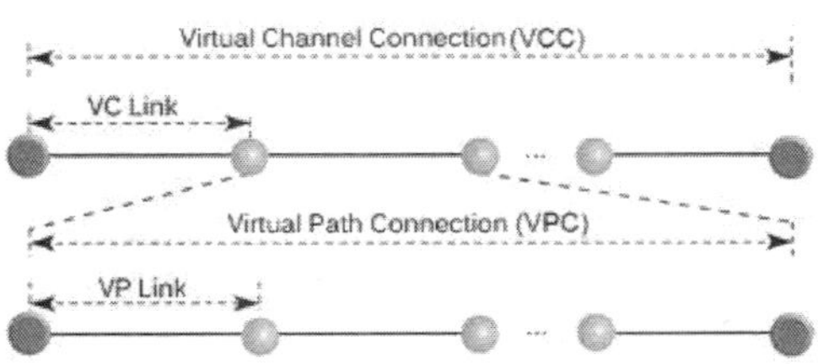

Link hierarchy VC, VCC, VP and VPC

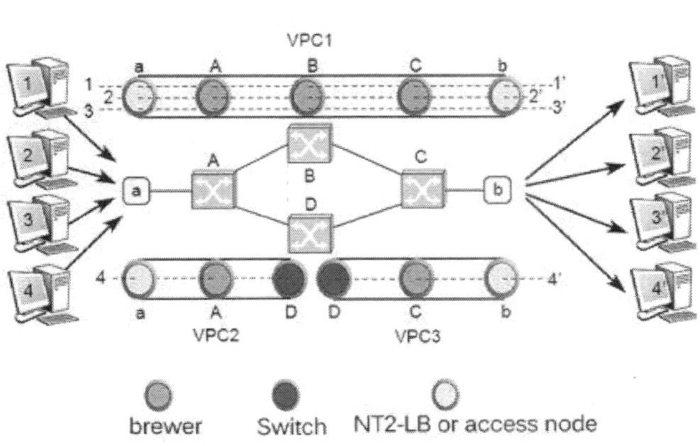

Fig VPC and VCC

3.4 ATM Networks

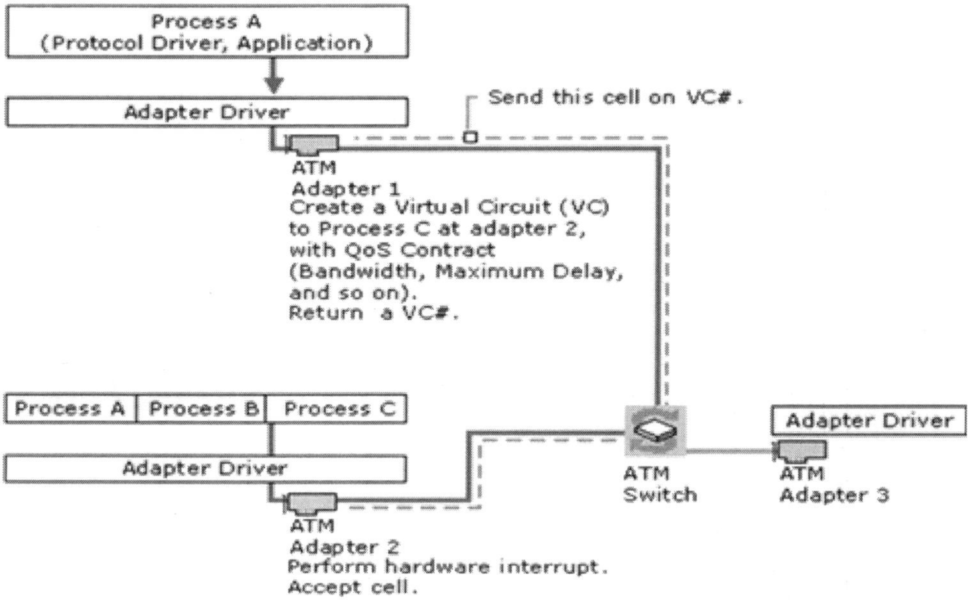

Fig ATM Network

- ATM contains two basic components: an endstation, a computer connected to the ATM network; and an ATM switch, the device responsible for connecting endstations and making sure data is transferred successfully.
- ATM is connection-oriented which establishes a defined path known as a virtual circuit (VC) sends a series of same-size frames called cells along the virtual circuit towards the destination.
- ATM endpoint negotiates a Quality of Service (QoS) contract for virtual circuit and spells out the bandwidth, maximum transit delay, acceptable variance in the transit delay, etc, that the VC provides, and this contract extends from one endpoint to the other through all of the intermediate ATM switches.
- ATM traffic is established at the outset, and the switching hardware merely needs to examine a simple header to identify the proper path.
- ATM allows a location to establish a full duplex connection with multiple locations at the same time.
- To travel across the ATM network, data is segmented into same-size cells, and encapsulated with a header that contains switching, congestion, and error-checking information.
- Cells are transmitted in order, and the ATM network uses Virtual Path Identifier and Virtual Channel Identifier (VPI/VCI) numbers in the ATM header to forward them efficiently.

❖ A switch reads the header, compares the VPI/VCI to its switching table to determine the correct output port and new VPI/VCI, and then forwards the cell.

❖ All of the addressing information the ATM switch needs is contained in the header and is always found in the same place. This makes the forwarding task simple to implement in hardware, reducing latency.

3.5 Software

With the migration to commodity PC hardware, standard commercial "off-the-shelf" operating systems and programming environments can be used inside of ATMs. Typical platforms previously used in ATM development include RMX or OS/2. Today the vast majority of ATMs worldwide use a Microsoft OS, primarily Windows XP Professional or Windows XP Embedded, a small number of deployments may still be running older versions such as Windows NT, Windows CE or Windows 2000. Notably, Vista was not widely adopted in ATMs, and as at October 2009 no bank has announced a Windows 7 deployment. Java, Linux and Unix may also be used in these environments, but such deployments are rare.

Linux is also finding some reception in the ATM marketplace. An example of this is Banrisul, the largest bank in the south of Brazil, which has replaced the MS-DOS operating systems in its ATMs with Linux. Banco do Brasil is also migrating ATMs to Linux.

Common application layer transaction protocols, such as Diebold 911 or 912, IBM PBM, and NCR NDC or NDC+ provide emulation of older generations of hardware on newer platforms with incremental extensions made over time to address new capabilities, although companies like NCR continuously improve these protocols issuing newer versions (latest NCR Aptra Advance NDC Version 3.x.y (Where x.y are subversions). Most major ATM manufacturers provide software packages that implement these protocols. Newer protocols such as IFX have yet to find wide acceptance by transaction processors.

With the move to a more standardized software base, financial institutions have been increasingly interested in the ability to pick and choose the application programs that drive their equipment. WOSA/XFS, now known as CEN XFS (or simply XFS), provides a common API for accessing and manipulating the various devices of an ATM.

J/XFS is a Java implementation of the CEN XFS API.

While the perceived benefit of XFS is similar to the Java's "Write once, run anywhere" mantra, often different ATM hardware vendors have different interpretations of the XFS standard. The result of these differences in interpretation means that ATM applications typically use a middleware to even out the differences between various platforms.

With the onset of Windows operating systems and XFS on ATM's, the software applications have the ability to become more intelligent. This has created a new breed of ATM applications commonly referred to as programmable applications. These types of applications allows for an entirely new host of applications in which the ATM terminal can do more than only communicate with the ATM switch. It is now empowered to connect to other content servers and video banking systems.

Notable ATM software that operates on XFS platforms include Triton PRISM, Diebold Agilis, CR2 BankWorld, KAL Kalignite, NCR Corporation Aptra Edge, Phoenix Interactive VISTAatm, and Wincor Nixdorf Protopas.

With the move of ATMs to industry-standard computing environments, concern has risen about the integrity of the ATM's software stack.

3.5.1 Features about ATM Software

The software to be designed will control a simulated automated teller machine (ATM) having a magnetic stripe reader for reading an ATM card, a customer console (keyboard and display) for interaction with the customer, a slot for depositing envelopes, a dispenser for cash (in multiples of $20), a printer for printing customer receipts, and a key-operated switch to allow an operator to start or stop the machine. The ATM will communicate with the bank's computer over an appropriate communication link.

The ATM will service one customer at a time. A customer will be required to insert an ATM card and enter a personal identification number (PIN) - both of which will be sent to the bank for validation as part of each transaction. The customer will then be able to perform one or more transactions. The card will be retained in the machine until the customer indicates that he/she desires no further transactions, at which point it will be returned - except as noted below.

The ATM must be able to provide the following services to the customer:

1. A customer must be able to make a cash withdrawal from any suitable account linked to the card, in multiples of $20.00. Approval must be obtained from the bank before cash is dispensed.

2. A customer must be able to make a deposit to any account linked to the card, consisting of cash and/or checks in an envelope. The customer will enter the amount of the deposit into the ATM, subject to manual verification when the envelope is removed from the machine by an operator. Approval must be obtained from the bank before physically accepting the envelope.

3. A customer must be able to make a transfer of money between any two accounts linked to the card.

4. A customer must be able to make a balance inquiry of any account linked to the card.

A customer must be able to abort a transaction in progress by pressing the Cancel key instead of responding to a request from the machine.

The ATM will communicate each transaction to the bank and obtain verification that it was allowed by the bank. Ordinarily, a transaction will be considered complete by the bank once it has been approved. In the case of a deposit, a second message will be sent to the bank indicating that the customer has deposited the envelope. (If the customer fails to deposit the envelope within the timeout period, or presses cancel instead, no second message will be sent to the bank and the deposit will not be credited to the customer.)

If the bank determines that the customer's PIN is invalid, the customer will be required to re-enter the PIN before a transaction can proceed. If the customer is unable to successfully enter the PIN after three tries, the card will be permanently retained by the machine, and the customer will have to contact the bank to get it back.

If a transaction fails for any reason other than an invalid PIN, the ATM will display an explanation of the problem, and will then ask the customer whether he/she wants to do another transaction.

The ATM will provide the customer with a printed receipt for each successful transaction, showing the date, time, machine location, type of transaction, account(s), amount, and ending and available balance(s) of the affected account ("to" account for transfers).

The ATM will have a key-operated switch that will allow an operator to start and stop the servicing of customers. After turning the switch to the "on" position, the operator will be required to verify and enter the total cash on hand. The machine can only be turned off when it is not servicing a customer. When the switch is moved to the "off" position, the machine will shut down, so that the operator may remove deposit envelopes and reload the machine with cash, blank receipts, etc.

The ATM will also maintain an internal log of transactions to facilitate resolving ambiguities arising from a hardware failure in the middle of a transaction. Entries will be made in the log when the ATM is started up and shut down, for each message sent to the Bank (along with the response back, if one is expected), for the dispensing of cash, and for the receiving of an envelope. Log entries may contain card numbers and dollar amounts, but for security will *never* contain a PIN.

System Startup

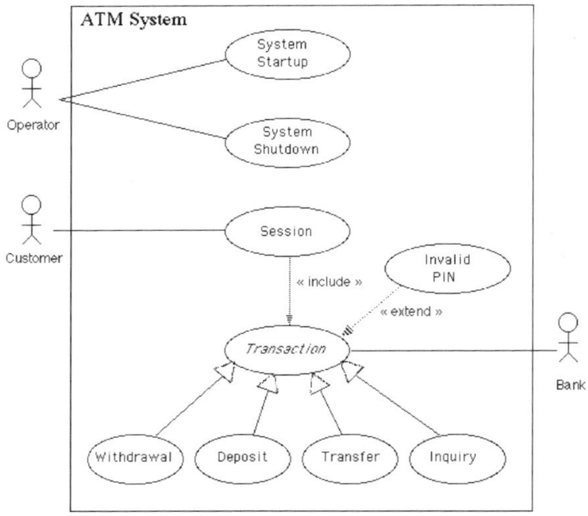

3.6 ATM System Design

Here we look at the design of the proposed ATM 'best of breed' menu system in relation to the potential users who could use the system. Figure shows a sequence diagram for a complete operational ATM system. The proposed 'best of breed' ATM system does not need to worry about factors such as, insufficient cash or invalid card, as it only concentrates on simulating an ATM navigation menu system.

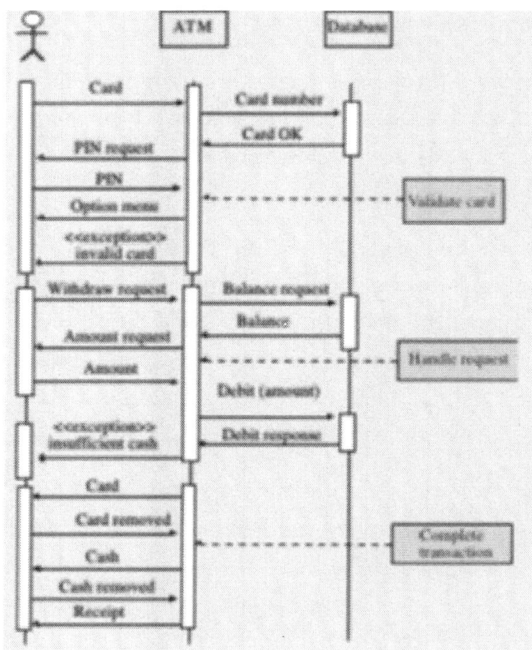

Fig: Sequence Diagram of ATM System

Existing ATMs menus will need to be mapped out. This is done by visiting each ATM and using the ATM, while at the same time drawing out the menu systems. To speed up the process of capturing the ATM menus a template was used as illustrated in Figure.

Fig: ATM menu screen

Figure represents a standard ATM screen and selection keys enabling the ATM menus to be drawn out quickly and more importantly accurately. It is crucial that each ATM menu system is mapped out accurately as these correspond exactly to the implemented version of the menu systems on the PC. The aim of transferring the ATM menu systems onto a VB program is to simulate the use of the actual ATM systems. Therefore, the performance of each ATM can be determined. When the real world ATM menu systems were drawn out with the aid of the template illustrated in Figure, the menu structures needed to be designed before they could be implemented in VB.NET.

Once the menu structures were designed and the different levels determined, this made the implementation stage an easier process. In the diagrams/tree structures- each box represents a particular menu screen.

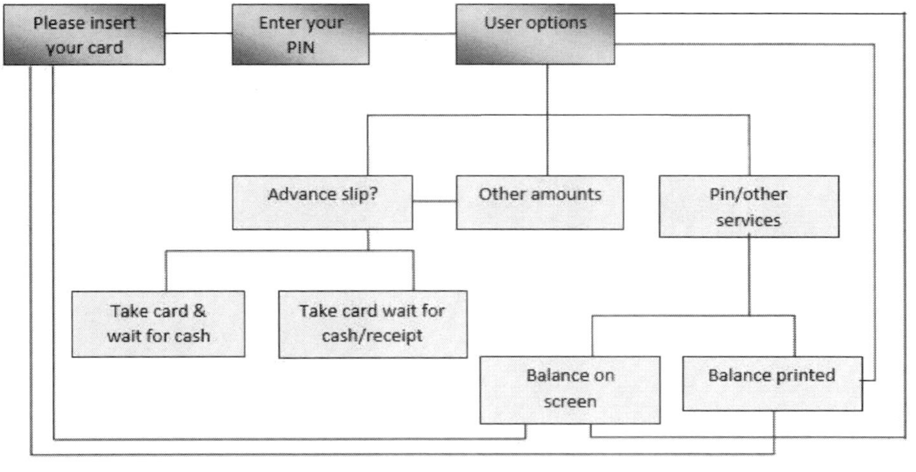

Fig: ATM Menu Tree Structure

The user can only move onto another menu screen after an input i.e. choosing an option. For example, a typical Bank ATM transaction may be:

- User inserts card as prompted
- User enters 4 digit PIN as prompted
- User opts to withdraw from menu selection
- The user opts to receive an advance slip when prompted
- The user is asked to take card and wait for cash and receipt.

To measure the transaction performance of the various ATMs, a VB program was created to simulate each ATM's real world menu structure to replicate ATM transactions in the lab. The ATM simulation program is used to test and monitor each of the ATMs performance. The user is presented with a collection of buttons. The user will click on the required button, taking the user to that particular banks ATM simulator.

The ATMs simulation user interface 'shell' remains consistent for each ATM i.e. the main screen, the selection keys and the keypad. This is illustrated in Figure. The eight selection keys (either side of the ATM screen) will be used to make user selections from the menu. The 'Insert Card' button will be used to simulate the user inserting the ATM card into the machine. The keypad containing the digits 0-9 and the keys 'Cancel', 'Clear' and 'Enter' is standard for all the various types of ATMs investigated in this research. The button 'Back to Main Menu' is simply to take the user back to the main screen illustrated in Figure.

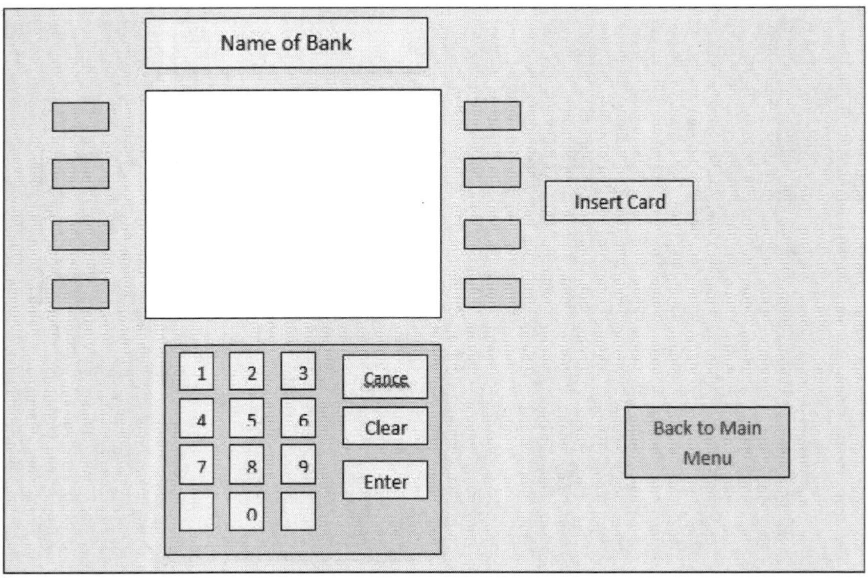

Fig ATM Simulator Interface Design

The 'Best of Breed' ATM menu system (called OptiATM which means optimal ATM menu) also uses the standard interface shown in Figure. The system is used to run transaction performance tests on the existing ATM menu designs (as well as the 'best-of-breed' OptiATM). When using an ATM the machine often takes time to process data such as 'processing card' and 'contacting bank/building society' etc. The ATM Simulator will not simulate these processing time periods. However, this will not corrupt the transaction performance test results as it will be consistent for all the ATM simulations. As these processing time periods will not be represented by any of the ATM simulations, the transaction performance test results will be accurate as they are all relative.

Another important factor to note is as follows; many inconveniences can occur when using an ATM, such as – ATM has run out of cash, user enters PIN incorrectly, error in reading card - to name a few. The ATM Simulator will not simulate these situations. When using the ATM simulator, 'perfect' transactions will be simulated i.e. the ATM reads the users card without error, the user enters the correct PIN and the user has sufficient funds. The ATM Simulator has a welcoming screen allowing the user to select which ATM to simulate. This is illustrated in Figure. Each of the Bank logos are buttons. The user simply clicks on a particular button to go to simulate that particular bank. At this stage, all the buttons take the user to all the existing ATM simulations.

Fig: ATM Simulator Main Screen

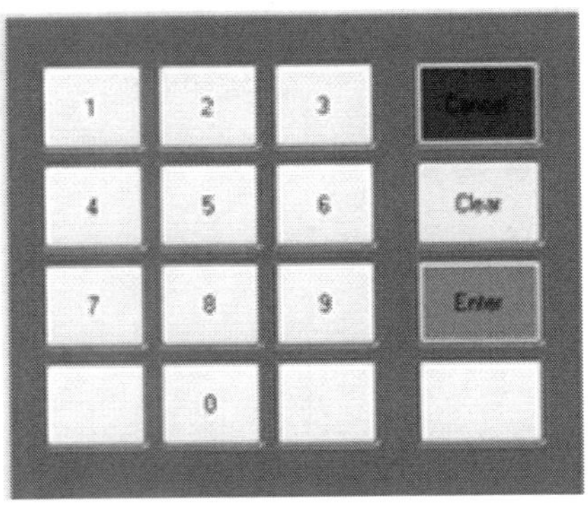

Fig: ATM keypad

Obviously it is not possible to physically insert an ATM card when using the ATM Simulator. Therefore an 'insert card' button has been created to simulate inserting the card into the ATM. Once this button is clicked by the user, the next menu screen appears. This of course, is the screen which prompts the user to enter their PIN (Personal Identification Number). The requirements of the ATM Simulator were simply to simulate the use of an ATM machine. Therefore this is what the ATM Simulator does. When the users enter their PIN, there are no comparison algorithms or checks to confirm that the PIN entered was indeed correct. As mentioned earlier, the ATM Simulator is just a tool to enable transaction performance tests to be conducted on ATM menu designs.

To enter a PIN, the user simply clicks four digits on the keypad shown in Figure. The PIN can be any four digit number so long as it ends with the digit '1'. When this number is clicked, a click event is triggered which takes the user to the next screen. The other two ATM designs (First Trust and Nationwide) require the user to press the 'enter' key to confirm the PIN entered is correct. In this case, the click event is triggered when the 'enter' key is clicked in the ATM Simulator. These two different actions are reflected in the real world ATM designs.

To select an option from the menu, the user simply clicks on the select key adjacent to the menu option displayed on screen. This is the same when using an ATM in real-life, only the user presses the selection key with their finger. An example of a user selection is illustrated in Fig. An Ulster Bank ATM user may want to withdraw cash. Therefore, once the user simulates inserting their card and entering their PIN, the user will click on the selection button adjacent to the option 'Withdraw cash'. This will then present the user with different cash withdrawal options.

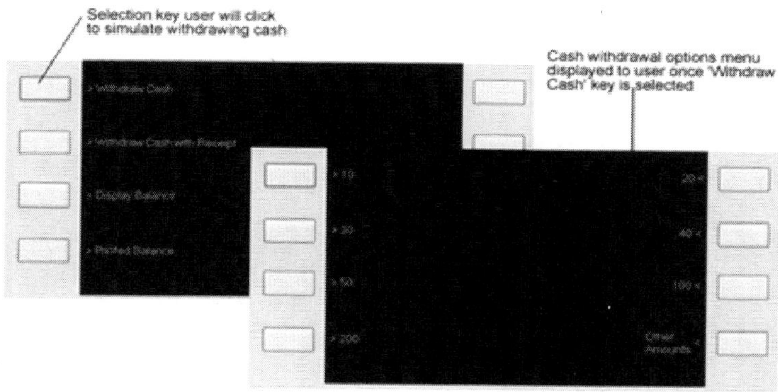

Fig: ATM - Withdraw Cash

Each existing ATM has been designed and implemented to reflect the real-life counterparts. Therefore each ATM, as illustrated in the design chapter using the

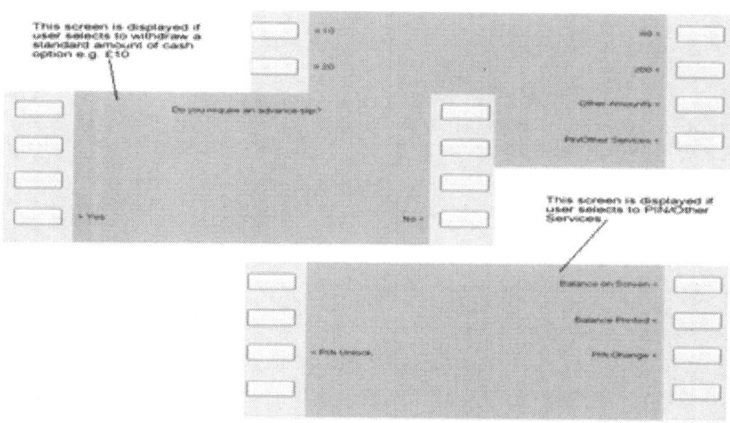

Fig: ATM User Options

ATM menu tree structures, will have different menus presenting the user with different options. Using the Bank of Ireland ATM, when the user enters their PIN, they are automatically presented with cash withdrawal options as well as some additional options. This is illustrated in Fig. Using the Ulster Bank ATM to withdraw cash was already illustrated in Fig. Using the Northern Bank ATM system, simulating user's options is illustrated in Fig. The initial menu shown here is only of course displayed to the user when the insertion of the ATM card and PIN entered is simulated previously.

As you can see from Fig, the user has three options when using Bank ATM i.e. withdraw cash, withdraw cash with a receipt and display or print balance.

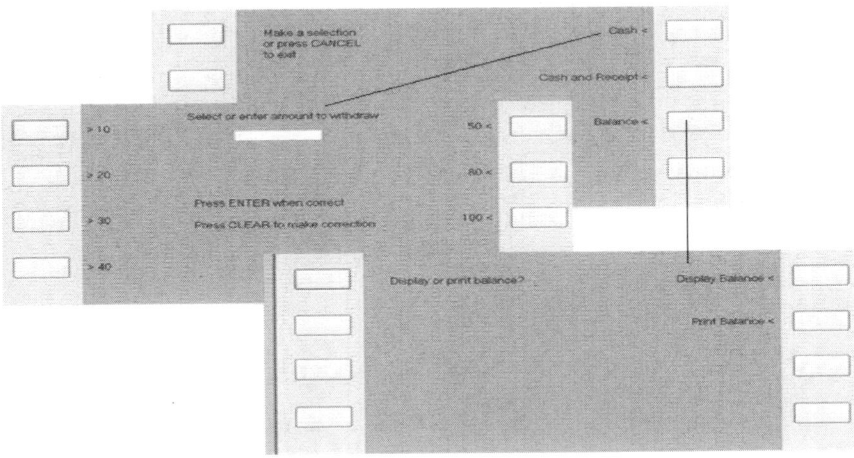

Fig: Bank ATM User Options

Figure show the options a user has when using the Nationwide ATM. Once the user enters their ATM card and PIN number, they are presented with the options; request statement, balance enquiry and cash withdrawal. Figure also shows the subsequent corresponding screens.

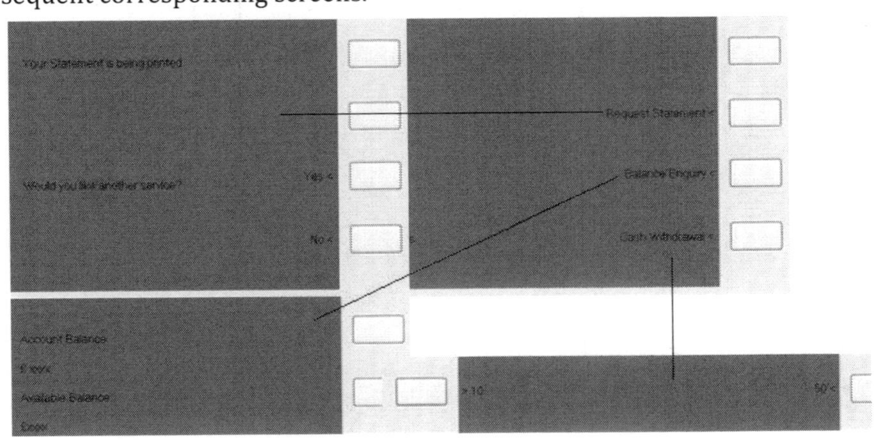

Fig: ATM User Options

First Trust also had its own unique menu design and layout. It was essential that each of the ATMs investigated (i.e. Bank of Ireland, Ulster Bank, First Trust, Nationwide and Northern Bank) be implemented correctly. It was essential that each ATM simulation on the PC directly represented the corresponding banks real life ATM menu design and layout to produce accurate and reliable transaction performance test results.

3.7 Events

3.7.1 System Startup Use Case

The system is started up when the operator turns the operator switch to the "on" position. The operator will be asked to enter the amount of money currently in the cash dispenser, and a connection to the bank will be established. Then the servicing of customers can begin.

System Startup Sequence Diagram

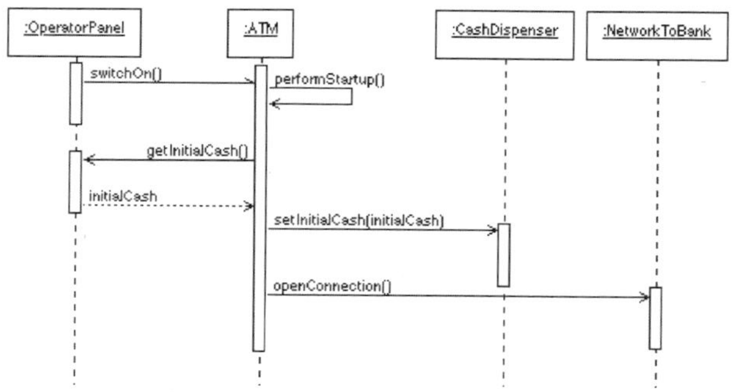

3.7.2 System Shutdown

The system is shut down when the operator makes sure that no customer is using the machine, and then turns the operator switch to the "off" position. The connection to the bank will be shut down. Then the operator is free to remove deposited envelopes, replenish cash and paper, etc.

System Shutdown Sequence Diagram

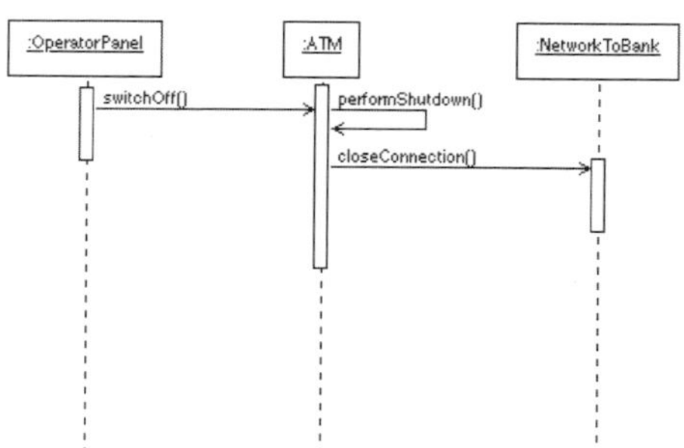

3.7.3 Session

A session is started when a customer inserts an ATM card into the card reader slot of the machine. The ATM pulls the card into the machine and reads it. (If the reader cannot read the card due to improper insertion or a damaged stripe, the card is ejected, an error screen is displayed, and the session is aborted.) The customer is asked to enter his/her PIN, and is then allowed to perform one or more transactions, choosing from a menu of possible types of transaction in each case. After each transaction, the customer is asked whether he/she would like to perform another. When the customer is through performing transactions, the card is ejected from the machine and the session ends. If a transaction is aborted due to too many invalid PIN entries, the session is also aborted, with the card being retained in the machine.

The customer may abort the session by pressing the Cancel key when entering a PIN or choosing a transaction type.

Session Sequence Diagram

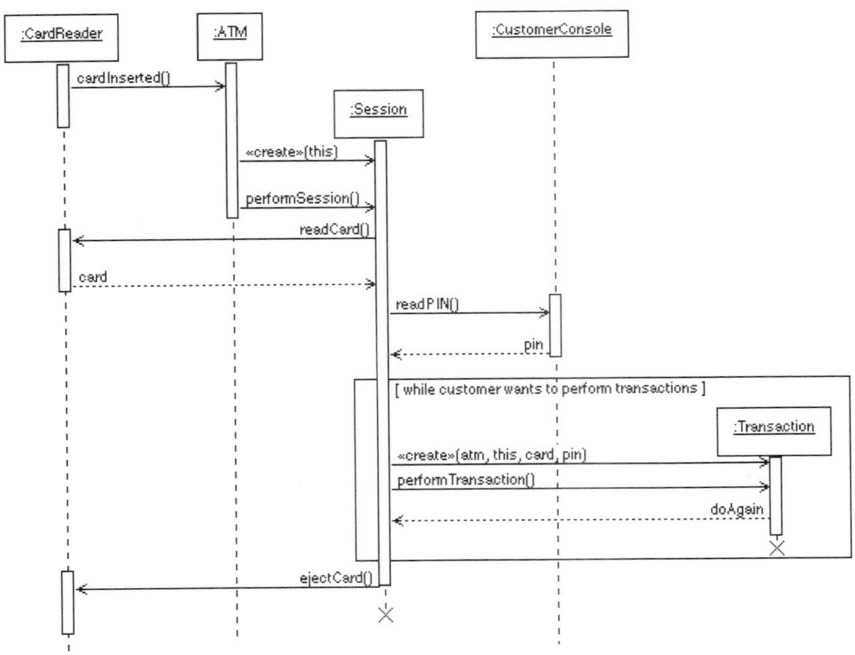

3.7.4 Transaction

Transaction is an abstract generalization. Each specific concrete type of transaction implements certain operations in the appropriate way. The flow of events given here describes the behavior common to all types of transaction. The flows of events for the individual types of transaction (withdrawal, deposit, transfer, inquiry) give the features that are specific to that type of transaction.

A transaction use case is started within a session when the customer chooses a transaction type from a menu of options. The customer will be asked to furnish appropriate details (e.g. account(s) involved, amount). The transaction will then be sent to the bank, along with information from the customer's card and the PIN the customer entered.

If the bank approves the transaction, any steps needed to complete the transaction (e.g. dispensing cash or accepting an envelope) will be performed, and then a receipt will be printed. Then the customer will be asked whether he/she wishes to do another transaction.

If the bank reports that the customer's PIN is invalid, the Invalid PIN extension will be performed and then an attempt will be made to continue the transaction. If the customer's card is retained due to too many invalid PINs, the transaction will be aborted, and the customer will not be offered the option of doing another.

If a transaction is cancelled by the customer, or fails for any reason other than repeated entries of an invalid PIN, a screen will be displayed informing the customer of the reason for the failure of the transaction, and then the customer will be offered the opportunity to do another.

The customer may cancel a transaction by pressing the Cancel key as described for each individual type of transaction below. All messages to the bank and responses back are recorded in the ATM's log.

Transaction Sequence Diagram

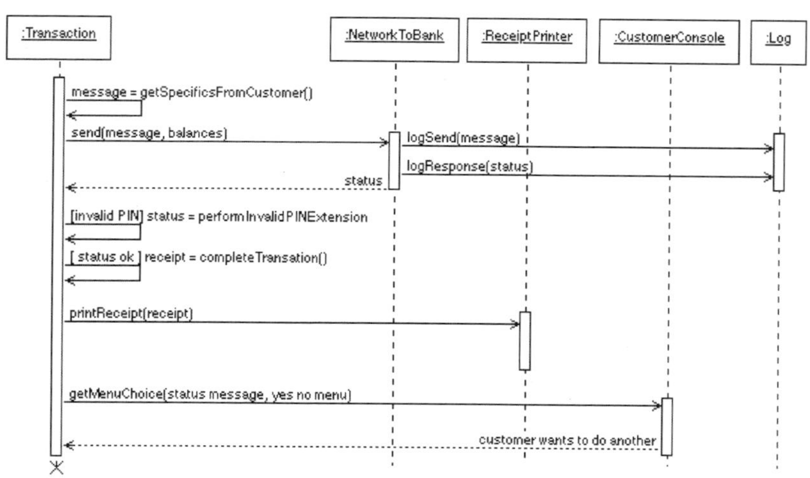

3.7.5 Withdrawal Transaction

A withdrawal transaction asks the customer to choose a type of account to withdraw from (e.g. checking) from a menu of possible accounts, and to choose a dollar amount from a menu of possible amounts. The system verifies that it has sufficient money on hand to satisfy the request before sending the transaction to the bank. (If not, the customer is informed and asked to enter a different amount.) If the transaction is approved by the bank, the appropriate amount of cash is dispensed by the machine before it issues a receipt. (The dispensing of cash is also recorded in the ATM's log.)

A withdrawal transaction can be cancelled by the customer pressing the Cancel key any time prior to choosing the dollar amount.

Withdrawal Transaction Collaboration

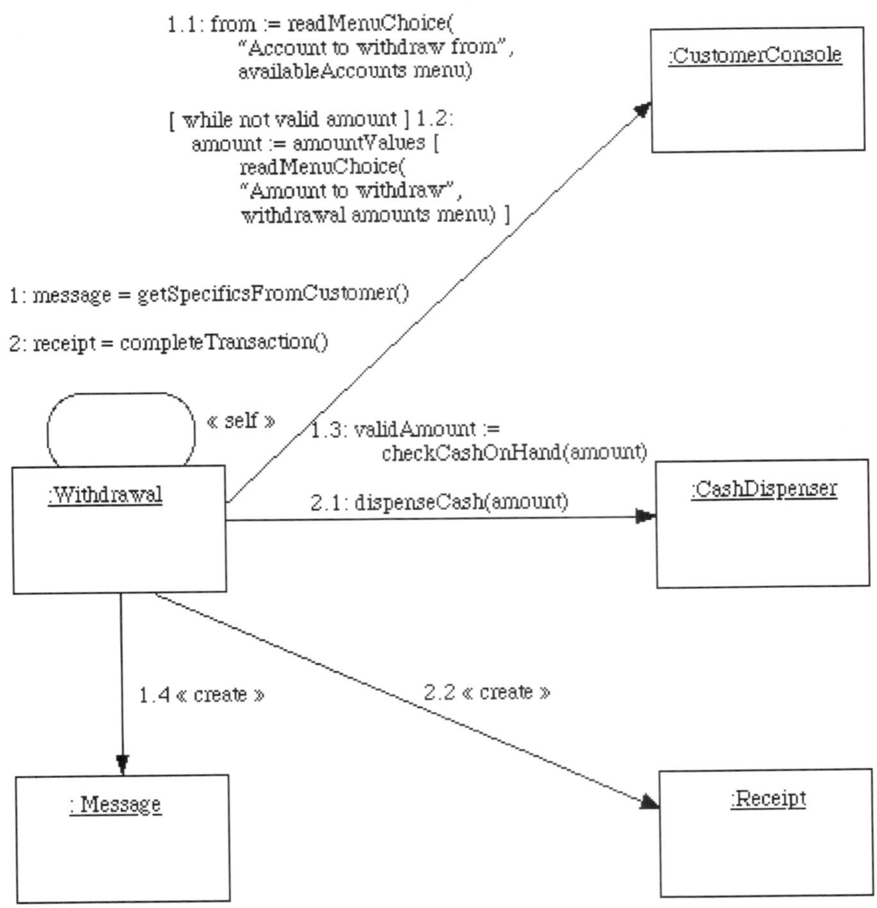

3.7.6 Deposit Transaction

A deposit transaction asks the customer to choose a type of account to deposit to (e.g. checking) from a menu of possible accounts, and to type in a dollar amount on the keyboard. The transaction is initially sent to the bank to verify that the ATM can accept a deposit from this customer to this account. If the transaction is approved, the machine accepts an envelope from the customer containing cash and/or checks before it issues a receipt. Once the envelope has been received, a second message is sent to the bank, to confirm that the bank can credit the customer's account - contingent on manual verification of the deposit envelope contents by an operator later. (The receipt of an envelope is also recorded in the ATM's log.)

A deposit transaction can be cancelled by the customer pressing the Cancel key any time prior to inserting the envelope containing the deposit. The transaction is automatically cancelled if the customer fails to insert the envelope containing the deposit within a reasonable period of time after being asked to do so.

Deposit Transaction Collaboration

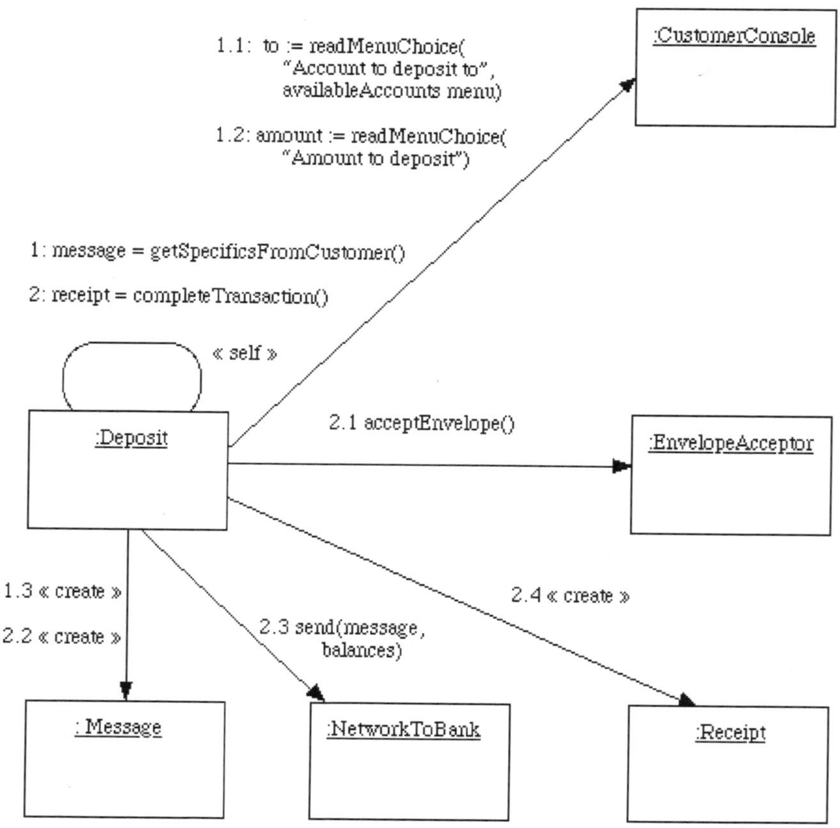

3.7.7 Transfer Transaction

A transfer transaction asks the customer to choose a type of account to transfer from (e.g. checking) from a menu of possible accounts, to choose a different account to transfer to, and to type in a dollar amount on the keyboard. No further action is required once the transaction is approved by the bank before printing the receipt.

A transfer transaction can be cancelled by the customer pressing the Cancel key any time prior to entering a dollar amount.

Transfer Transaction Collaboration

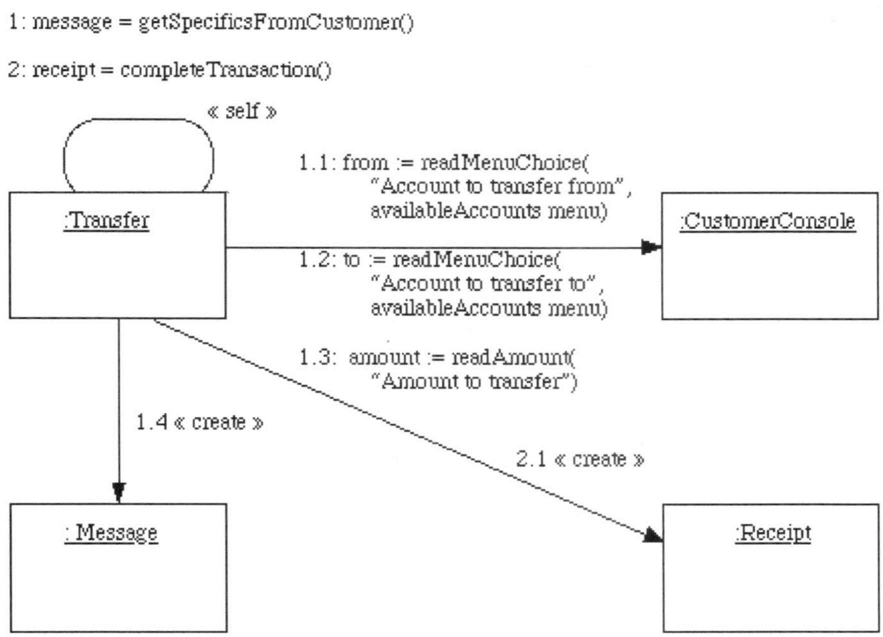

3.7.8 Inquiry Transaction

An inquiry transaction asks the customer to choose a type of account to inquire about from a menu of possible accounts. No further action is required once the transaction is approved by the bank before printing the receipt.

An inquiry transaction can be cancelled by the customer pressing the Cancel key any time prior to choosing the account to inquire about.

Inquiry Transaction Collaboration

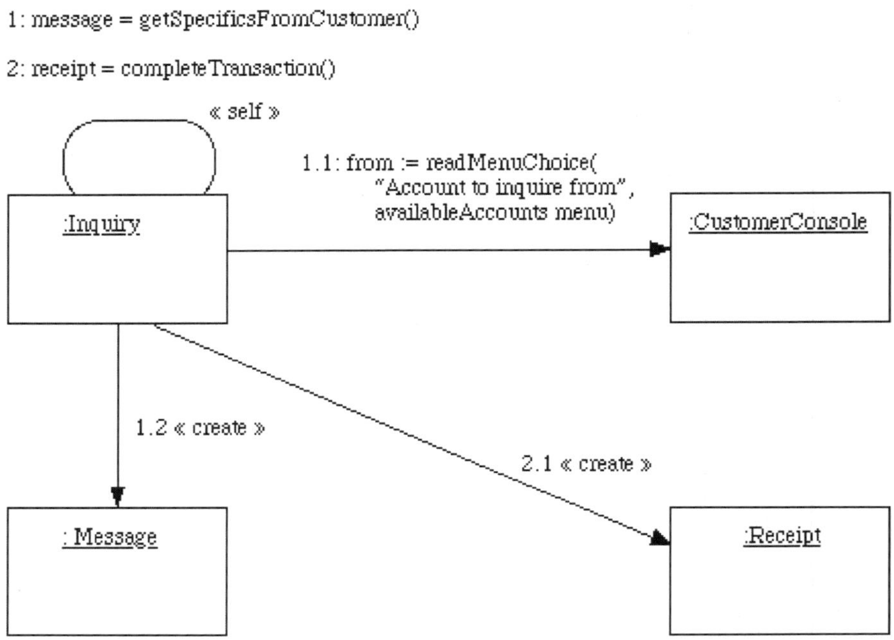

3.7.9 Invalid PIN Extension

An invalid PIN extension is started from within a transaction when the bank reports that the customer's transaction is disapproved due to an invalid PIN. The customer is required to re-enter the PIN and the original request is sent to the bank again. If the bank now approves the transaction, or disapproves it for some other reason, the original use case is continued; otherwise the process of re-entering the PIN is repeated. Once the PIN is successfully re-entered, it is used for both the current transaction and all subsequent transactions in the session. If the customer fails three times to enter the correct PIN, the card is permanently retained, a screen is displayed informing the customer of this and suggesting he/she contact the bank, and the entire customer session is aborted.

If the customer presses Cancel instead of re-entering a PIN, the original transaction is cancelled.

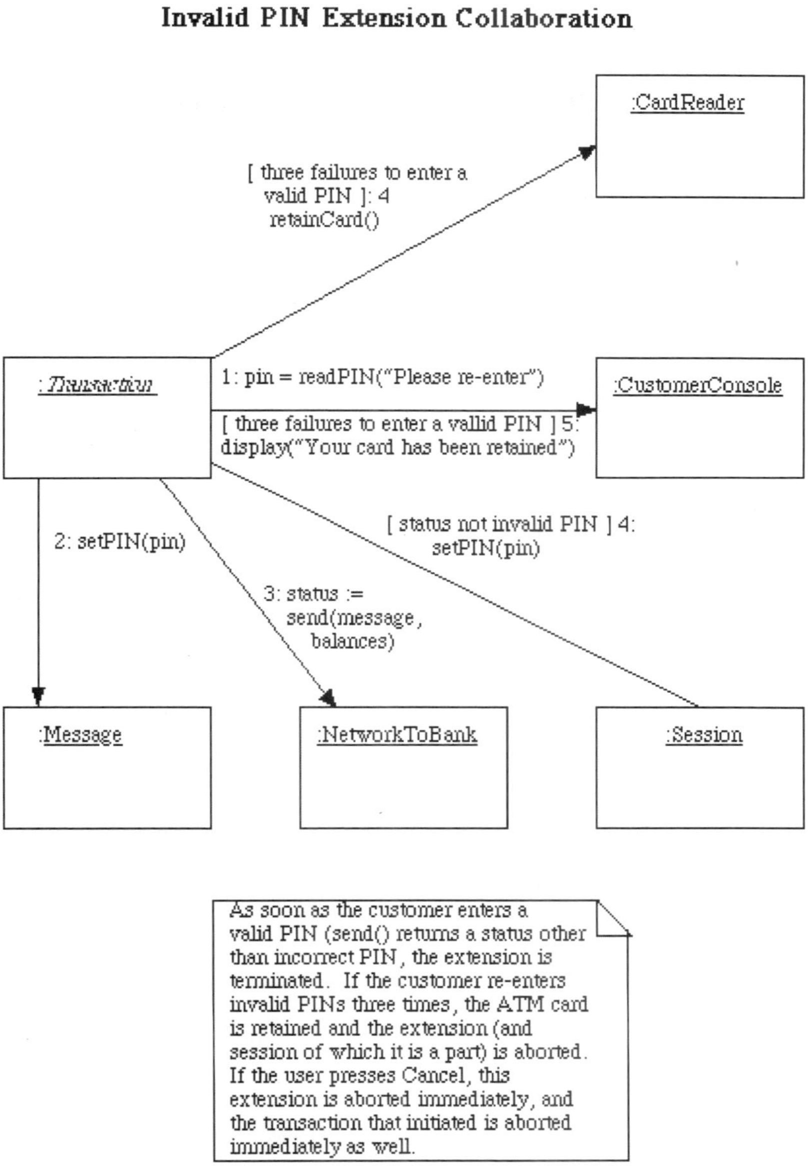

3.8 Testing and Errors

3.8.1 Testing

Use Case	Function Being Tested	Initial System State	Input	Expected Output
System Startup	System is started when the switch is turned "on"	System is off	Activate the "on" switch	System requests initial cash amount

Use Case	Function Being Tested	Initial System State	Input	Expected Output
System Startup	System accepts initial cash amount	System is requesting cash amount	Enter a legitimate amount	System is on
System Startup	Connection to the bank is established	System has just been turned on	Perform a legitimate inquiry transaction	System output should demonstrate that a connection has been established to the Bank
System Shutdown	System is shut down when the switch is turned "off"	System is on and not servicing a customer	Activate the "off" switch	System is off
System Shutdown	Connection to the Bank is terminated when the system is shut down	System has just been shut down		Verify from the bank side that a connection to the ATM no longer exists
Session	System reads a customer's ATM card	System is on and not servicing a customer	Insert a readable card	Card is accepted; System asks for entry of PIN
Session	System rejects an unreadable card	System is on and not servicing a customer	Insert an unreadable card	Card is ejected; System displays an error screen; System is ready to start a new session
Session	System accepts customer's PIN	System is asking for entry of PIN	Enter a PIN	System displays a menu of transaction types
Session	System allows customer to perform a transaction	System is displaying menu of transaction types	Perform a transaction	System asks whether customer wants another transaction
Session	System allows multiple transactions in one session	System is asking whether customer wants another transaction	Answer yes	System displays a menu of transaction types
Session	Session ends when customer chooses not to do another transaction	System is asking whether customer wants another transaction	Answer no	System ejects card and is ready to start a new session

Use Case	Function Being Tested	Initial System State	Input	Expected Output
Transaction	Individual types of transaction will be tested below			
Transaction	System handles an invalid PIN properly	A readable card has been entered	Enter an incorrect PIN and then attempt a transaction	The Invalid PIN Extension is performed
Withdrawal	System asks customer to choose an account to withdraw from	Menu of transaction types is being displayed	Choose Withdrawal transaction	System displays a menu of account types
Withdrawal	System asks customer to choose a dollar amount to withdraw	Menu of account types is being displayed	Choose checking account	System displays a menu of possible withdrawal amounts
Withdrawal	System performs a legitimate withdrawal transaction properly	System is displaying the menu of withdrawal amounts	Choose an amount that the system currently has and which is not greater than the account balance	System dispenses this amount of cash; System prints a correct receipt showing amount and correct updated balance; System records transaction correctly in the log (showing both message to the bank and approval back)
Withdrawal	System verifies that it has sufficient cash on hand to fulfill the request	System has been started up with less than the maximum withdrawal amount in cash on hand; System is requesting a withdrawal amount	Choose an amount greater than what the system currently has	System displays an appropriate message and asks customer to choose a different amount
Withdrawal	System verifies that customer's balance is sufficient to fulfill the request	System is requesting a withdrawal ammount	Choose an amount that the system currently has but which is greater than the account balance	System displays an appropriate message and offers customer the option of choosing to do another transaction or not.

Use Case	Function Being Tested	Initial System State	Input	Expected Output
Withdrawal	A withdrawal transaction can be cancelled by the customer any time prior to choosing the dollar amount	System is displaying menu of account types	Press "Cancel" key	System displays an appropriate message and offers customer the option of choosing to do another transaction or not.
Withdrawal	A withdrawal transaction can be cancelled by the customer any time prior to choosing the dollar amount	System is displaying menu of dollar amounts	Press "Cancel" key	System displays an appropriate message and offers customer the option of choosing to do another transaction or not.
Deposit	System asks customer to choose an account to deposit to	Menu of transaction types is being displayed	Choose Deposit transaction	System displays a menu of account types
Deposit	System asks customer to enter a dollar amount to deposit	Menu of account types is being displayed	Choose checking account	System displays a request for the customer to type a dollar amount
Deposit	System asks customer to insert an envelope	System is displaying a request for the customer to type a dollar amount	Enter a legitimate dollar amount	System requests that customer insert an envelope
Deposit	System performs a legitimate deposit transaction properly	System is requesting that customer insert an envelope	Insert an envelope	System accepts envelope; System prints a correct receipt showing amount and correct updated balance; System records transaction correctly in the log (showing message to the bank, approval back, and acceptance of the envelope)
Deposit	A deposit transaction can be cancelled by the customer any time prior to inserting an envelope	System is displaying menu of account types	Press "Cancel" key	System displays an appropriate message and offers customer the option of choosing to do another transaction or not.

Use Case	Function Being Tested	Initial System State	Input	Expected Output
Deposit	A deposit transaction can be cancelled by the customer any time prior to inserting an envelope	System is requesting customer to enter a dollar amount	Press "Cancel" key	System displays an appropriate message and offers customer the option of choosing to do another transaction or not.
Deposit	A deposit transaction can be cancelled by the customer any time prior to inserting an envelope	System is requesting customer to insert an envelope	Press "Cancel" key	System displays an appropriate message and offers customer the option of choosing to do another transaction or not.
Deposit	A deposit transaction is cancelled automatically if an envelope is not inserted within a reasonable time	System is requesting customer to insert an envelope	Wait for the request to time out	System displays an appropriate message and offers customer the option of choosing to do another transaction or not.
Transfer	System asks customer to choose an account to transfer from	Menu of transaction types is being displayed	Choose Transfer transaction	System displays a menu of account types specifying transfer from
Transfer	System asks customer to choose an account to transfer to	Menu of account types to transfer from is being displayed	Choose checking account	System displays a menu of account types specifying transfer to
Transfer	System asks customer to enter a dollar amount to transfer	Menu of account types to transfer to is being displayed	Choose savings account	System displays a request for the customer to type a dollar amount
Transfer	System performs a legitimate transfer transaction properly	System is displaying a request for the customer to type a dollar amount	Enter a legitimate dollar amount	System prints a correct receipt showing amount and correct updated balance; System records transaction correctly in the log (showing both message to the bank and approval back)

Use Case	Function Being Tested	Initial System State	Input	Expected Output
Transfer	A transfer transaction can be cancelled by the customer any time prior to entering dollar amount	System is displaying menu of account types specifying transfer from	Press "Cancel" key	System displays an appropriate message and offers customer the option of choosing to do another transaction or not.
Transfer	A transfer transaction can be cancelled by the customer any time prior to entering dollar amount	System is displaying menu of account types specifying transfer to	Press "Cancel" key	System displays an appropriate message and offers customer the option of choosing to do another transaction or not.
Transfer	A transfer transaction can be cancelled by the customer any time prior to entering dollar amount	System is requesting customer to enter a dollar amount	Press "Cancel" key	System displays an appropriate message and offers customer the option of choosing to do another transaction or not.
Inquiry	System asks customer to choose an account to inquire about	Menu of transaction types is being displayed	Choose Inquiry transaction	System displays a menu of account types
Inquiry	System performs a legitimate inquiry transaction properly	System is displaying menu of account types	Choose checking account	System prints a correct receipt showing correct balance; System records transaction correctly in the log (showing both message to the bank and approval back)
Inquiry	An inquiry transaction can be cancelled by the customer any time prior to choosing an account	System is displaying menu of account types	Press "Cancel" key	System displays an appropriate message and offers customer the option of choosing to do another transaction or not.

Use Case	Function Being Tested	Initial System State	Input	Expected Output
Invalid PIN Extension	Customer is asked to reenter PIN		Enter an incorrect PIN; Attempt an inquiry transaction on the customer's checking account	Customer is asked to re-enter PIN
Invalid PIN Extension	Correct re-entry of PIN is accepted	Request to re-enter PIN is being displayed	Enter correct PIN	Original transaction completes successfully
Invalid PIN Extension	A correctly re-entered PIN is used for subsequent transactions	An incorrect PIN has been re-entered and transaction completed normally	Perform another transaction	This transaction completes successfully as well
Invalid PIN Extension	Incorrect re-entry of PIN is not accepted	Request to re-enter PIN is being displayed	Enter incorrect PIN	An appropriate message is displayed and re-entry of the PIN is again requested
Invalid PIN Extension	Correct re-entry of PIN on the second try is accepted	Request to re-enter PIN is being displayed	Enter incorrect PIN the first time, then correct PIN the second time	Original transaction completes successfully
Invalid PIN Extension	Correct re-entry of PIN on the third try is accepted	Request to re-enter PIN is being displayed	Enter incorrect PIN the first time and second times, then correct PIN the third time	Original transaction completes successfully
Invalid PIN Extension	Three incorrect re-entries of PIN result in retaining card and aborting transaction	Request to re-enter PIN is being displayed	Enter incorrect PIN three times	An appropriate message is displayed; Card is retained by machine; Session is terminated

3.8.2 Errors

Error 0110091: Lost Card & Slip

Solution: Remove the retracted card and receipt in each retract box. Then execute reset operation.

Error 0110291: Lost withdraw cash

Solution: Remove the retracted cash in cash dispenser retract box, then execute Cassette Total operation.

Error 0110391: Lost Card & Slip

Solution: Remove the retracted card and receipt in IPM retract box. Then execute IPM Cassette Total operation.

Error 0B00001: The USR_Bill_Inf_Stacker_Status Is Normal

Error 0B00002: Note_Status of Bill is not "Notpresent"

Error 0B00003: Total_Bill_Count is over than the APP-Full_Bill_Stacker

Error 0B00010: Device_Status in "NO Device"

Error 0D00001: The Status of Door is opened

Solution: Close the door

Error 0N00010: Status of Network is not normal

Error 1: TIME-OUT

Cause:

1. Request has been sent. Time-out waiting for response or carrier was dropped. No characters were received. Time-out is typically 60 seconds.
2. Request has been sent. Time-out waiting for response or carrier was dropped. At least one character was received (ETX was never received). Time-out is typically 60 seconds.
3. EOT was received as first character after sending a NAK due to a bad response message.

Result:

No money is dispensed, screen and receipt display system unavailable.

Error 101: SECOND TIME-OUT WAITING FOR PICK (FEED FAILURE)

How to fix the problem:

1. Refill the cassette as needed.
2. Inspect the feed path for currency that is stuck together or jammed. If no jams are found, remove the note closest to the pick rollers in the cassette.
3. Install the cassette. Purge the dispensing mechanism with the PURGE command from the DIAGNOSTICS function. Test the TDM-100 by completing several Test Dispenses. If the error clears, put the ATM in service.
4. If the currency is in "fit" condition and the error condition persists, consider replacing the note cassette or the dispensing mechanism.

Error 102: ERROR (PICK MOTOR OVER CURRENT)

How to fix the problem:

1. Inspect the cassette and feed path for jammed currency. Clear any jammed currency.
2. Verify that all access panels are closed and secured. Purge the dispenser using the PURGE command.
3. Test the dispensing mechanism by completing several Test Dispenses. If the error does not occur again, put the ATM in service.
4. If the error persists, replace the dispensing mechanism.

Error 103: THICKNESS SENSOR UNSTABLE

Enter the diagnostic function and complete the "learn note thickness" command. Clear the error. Purge the dispenser using the PURGE command. Test the dispensing mechanism by completing several test dispenses. If the error does not occur again, put the ATM in service. If the error persists, replace the dispensing mechanism.

Error 1030100: DEV_PIN (Key data error from host).

Contact host.

Error 104: UNABLE TO CLEAR WIDTH SENSOR

fix the problem:

1. Remove the cassette and inspect for jammed currency in the width sensor and at the output of the cassette. The width sensor may be dirty. Clean the width sensor with compressed air. Purge the dispenser using the PURGE command. Test the dispensing mechanism by completing several test dispenses. If the error does not occur again, put the ATM in service. If the error persists, it may be necessary to replace either the cassette or dispensing mechanism.

Error 104: Unable to clear width sensor during reject

Error 105: Triton ATM and/or Hyosung ATM Standard 3 Error Code

There may be an insufficient number of notes in the cassette to complete the requested amount (minimum number of 15 notes). Press the "learn note thickness" command. Put more currency in the cassette and repeat the "learn note thickness" command.

Error 106: FIFO ERROR

fix the problem:

1. The dispensing mechanism may have corrupt software. Power down the ATM in the proper sequence. From the Main Menu screen, select System Parameters (5) followed by Shut Down the Terminal (4). When the indication that it is safe to turn off the terminal is displayed on the screen, push the Power ON/OFF switch to the OFF position. After about ten seconds, push the Power ON/OFF switch to the ON position to start the terminal.
2. Purge the dispenser using the PURGE command. Test the dispensing mechanism by completing several test dispenses.
3. If the error clears, put the ATM into service. If the error persists, replace the dispensing mechanism.

Error 107: Suspect reject accountancy

Error 108: UNEXPECTED NOTE AT DOUBLE DETECT

fix the problem:

1. Inspect the dispensing mechanism for damaged components or broken wires.
2. Remove the cassette and visually inspect for loose or disconnected connectors on both the upper and lower width sensor printed circuit boards.
3. If there is no visible damage and the connectors are attached to the sensor boards, the best course of action is to replace the dispensing mechanism.

Error 109: TIME-OUT AT EXIT SENSOR

fix the problem:

1. Inspect for jammed currency in the feed path and at the Exit sensor. Remove any jammed currency.
2. Purge the dispenser using the PURGE command. Test the dispensing mechanism by completing several Test Dispenses.
3. If the error clears, put the ATM into service. Otherwise, replace the dispensing mechanism.

Error 11:NO CONNECT

Cause:

1. Used up all radial attempts and got Busy Signal for each attempt or No Dial Tone for each attempt.
2. Got connected and never received ENQ within time-out period (14 seconds.)

Result:

No money is dispensed, screen and receipt display system unavailable.

Error 110: TRAILING EDGE TIME-OUT AT EXIT

fix the problem:

1. Inspect for jammed currency in the feed path and at the Exit sensor. Remove any jammed currency.
2. Purge the dispenser using the PURGE command.
3. Test the dispensing mechanism by completing several test dispenses.
4. If the error clears, put the ATM into service. If the error persists, replace the dispensing mechanism.

Error 1100910: Lost Slip

Solution: Check if the receipt printer is installed properly.

Error 1101910: LOST CARD", DEV_MCU

Error 1102910: LOST WITHDRAW CASH", DEV_CSH.

Error 1103910: LOST CARD & SLIP", DEV_MCU | DEV_SPR.

Error 1106910: LOST SLIP", DEV_SPR

Error 111: DIVERTER TIME-OUT

fix the problem:

1. Inspect for jammed currency at the diverter. Remove jammed currency.
2. Verify that all access panels are closed and secured. Clear the error. Test the dispenser by completing several test dispenses.
3. If the error clears, put the ATM in service. If these actions have no effect on the dispenser operation, replace the dispenser mechanism.

Error 111: Diverter Timeout

Error 112: TIME-OUT WAITING FOR LEADING EDGE AT REJECT

fix the problem:

1. Inspect for jammed currency in the transport path between the width sensors and the reject sensor. Remove jammed currency.
2. Verify that all access panels are closed and secured. Clear the error. Test the dispenser by completing several test dispenses.

3. If the error clears, put the ATM in service. If these actions have no effect on dispenser operation, replace the dispenser mechanism.

Error 112: Timeout waiting for leading edge at reject

Error 113: TIME-OUT WAITING FOR TRAILING EDGE AT REJECT

fix the problem:

1. Inspect for a jammed currency in the reject sensor.
2. Verify that all access panesl are closed and secured. Power down the ATM in the proper sequence. From the Main Menu screen, select System Parameters (5) followed by Shut Down the Terminal (4). When the indication that it is safe to turn off the terminal is displayed on the screen, push the Power ON/OFF switch to the OFF position. After about ten seconds, push the Power ON/OFF switch to the ON position to start the terminal.
3. Test the dispenser by completing several test dispenses.
4. If the error clears, put the ATM in service. If these actions have no effect on dispenser operation, replace the dispenser.

Error 114: EXIT BLOCKED DURING PURGE

fix the problem:

1. Inspect the dispensing mechanism for jammed currency at the exit sensor. Clear any jammed currency.
2. Clean the exit sensor using a soft brush and a vacuum cleaner. Power down the ATM in the proper sequence. From the Main Menu screen, select System Parameters (5) followed by Shut Down the Terminal (4). When the indication that it is safe to turn off the terminal is displayed on the screen, push the Power ON/OFF switch to the OFF position. After about ten seconds, push the Power ON/OFF switch to the ON position to start the terminal.
3. Clear the error.
4. Test the dispenser by completing several test dispenses.
5. If the error clears, put the ATM in service. If these actions have no effect on dispenser operation, replace the dispenser mechanism.

Error 115: DIVERTER TIME-OUT ON PURGE

fix the problem:

1. Inspect the dispensing mechanism for jammed currency at the diverter. Clear any jammed currency.

2. Make sure the shelf that the dispenser is mounted on is level and seated at all four corners.

3. Make sure the diverter moves freely.

4. Power down the ATM in the proper sequence. From the Main Menu screen, select System Parameters (5) followed by Shut Down the Terminal (4). When the indication that it is safe to turn off the terminal is displayed on the screen, push the Power ON/OFF switch to the OFF position. After about ten seconds, push the Power ON/OFF switch to the ON position to start the terminal.

5. Clear the error.

6. Test the dispenser by completing several test dispenses. If the error clears, put the ATM in service.

7. If these actions have no effect on dispenser operation, replace the dispenser mechanism.

Error 118: EXIT SENSOR BLOCKED ON START OF DISPENSE OR LEARN

fix the problem:

1. Inspect for jammed currency at the Exit. Remove jammed currency.

2. Verify that all access panels are closed and secured.

3. Use a soft brush and vacuum cleaner to clean the exit sensor.

4. Power down the ATM in the proper sequence. From the Main Menu screen, select System Parameters (5) followed by Shut Down the Terminal (4). When the indication that it is safe to turn off the terminal is displayed on the screen, push the Power ON/OFF switch to the OFF position. After about ten seconds, push the Power ON/OFF switch to the ON position to start the terminal.

5. Clear the error. Test the dispenser by completing several test dispenses.

6. If the error clears, put the ATM in service.

7. If these actions have no effect on dispenser operation, replace the dispenser mechanism.

Error 119: DIVERTER IN DISPENSE POSITION ON START OF DISPENSE OR LEARN

fix the problem:

1. Inspect for jammed currency at the Diverter. Remove jammed currency.

2. Check the operation of the diverter solenoid.
3. Verify that all access panels are closed and secured. Power down the ATM in the proper sequence. From the Main Menu screen, select System Parameters (5) followed by Shut Down the Terminal (4). When the indication that it is safe to turn off the terminal is displayed on the screen, push the Power ON/OFF switch to the OFF position. After about ten seconds, push the Power ON/OFF switch to the ON position to start the terminal.
4. Clear the error. Test the dispenser by completing several test dispenses.
5. If the error clears, put the ATM in service.
6. If these actions have no effect on dispenser operation, replace the dispenser mechanism.

Error 12: Stream Feed Channel

Error 120: Reject cassette not present

Error 121: Note cassette not present

Error 122:UNEXPECTED NOT AT EXIT

fix the problem:

1. Purge the dispenser using the PURGE command from the diagnostic menu.
2. Test the dispenser by completing several test dispenses. If the error clears, put the ATM in service.
3. If these actions have no effect on dispenser operation, replace the dispenser mechanism and/or the note cassette.

Error 124:DIVERTER MOVED TO EXIT POSITION DURING REJECT PURGE

fix the problem:

1. Inspect for a currency jam at the diverter.
2. Verify that the diverter moves freely.
3. Test the dispenser by completing several test dispenses.
4. If the error clears, put the ATM in service.
5. If these actions have no effect on dispenser operation, replace the dispenser mechanism.

Error 125: INITIAL STATUS CHECK FAILED

fix the problem:

1. Reset the ATM. Clear the error.

2. If the error persists, replace the dispensing mechanism.

Error 126: DIVERTER MOVED TO REJECT POSITION DURING DISPENSE

fix the problem:

1. Inspect for a note jam at the diverter. Verify that the diverter moves freely. Test the dispenser by completing several test dispenses. If the error clears, put the ATM in service. If these actions have no effect on dispenser operation, replace the dispenser mechanism.

Error 128: ERROR IN REPLY FROM THE DISPENSER MECHANISM

fix the problem:

1. Verify that the power supply is operating and the DC voltages being supplied to the main board and dispenser are correct. Reset connectors or replace the power supply as needed.

2. Verify that the main board to electronic journal and the electronic journal dispenser cables are securely seated at both ends of the cable. The cables may be defective.

3. Other components that may cause this error are the main board or the dispensing mechanism.

Error 13: BGD DONE GOOD

Cause:

Communication was normal, but response header does not match request.

Error 131: No CTS ready from dispenser

Error 132: Status reported bad double detect in previous dispense

Error 133: +5 VDC NOT PRESENT ON CARRIER DETECT

Error 1333900: No response form IC card

Error 134: EXIT BLOCKED AS REPORTED BY STATUS CHECK

fix the problem:

1. Inspect the feed path and exit sensor for jammed currency and broken components. The exit sensor may be dirty. Clean as needed with a soft brush and vacuum cleaner.

2. Power down the ATM in the proper sequence. From the Main Menu screen, select System Parameters (5) followed by Shut Down the Terminal (4).

When the indication that it is safe to turn off the terminal is displayed on the screen, push the Power ON/OFF switch to the OFF position. After about ten seconds, push the Power ON/OFF switch to the ON position to start the terminal.

3. Purge the dispenser with the PURGE command. Verify correct operation with a live transaction.
4. If the problem persists, replace the dispensing mechanism.

Error 135: FEED SENSOR BLOCKED AS REPORTED BY STATUS CHECK

fix the problem:

1. Inspect the feed path and feed sensors for jammed currency and broken components. The feed sensor may be dirty. Clean as needed with a soft brush and vacuum cleaner.
2. Power down the ATM in the proper sequence. From the Main Menu screen, select System Parameters (5) followed by Shut Down the Terminal (4). When the indication that it is safe to turn off the terminal is displayed on the screen, push the Power ON/OFF switch to the OFF position. After about ten seconds, push the Power ON/OFF switch to the ON position to start the terminal.
3. Purge the dispenser with the PURGE command.
4. Complete several test dispenses to verify correct operation. If the problem persists, replace the dispensing mechanism.

Error 136: MODEM INITIALIZATION FAILED

fix the problem:

1. Power down the ATM in the proper sequence. From the Main Menu screen, select System Parameters (5) followed by Shut Down the Terminal (4). When the indication that it is safe to turn off the terminal is displayed on the screen, push the Power ON/OFF switch to the OFF position. After about ten seconds, push the Power ON/OFF switch to the ON position to start the terminal.
2. If the problem persists, remove the main board and replace the modem.
3. If replacing the modem does not correct the problem, replace the main board.

Error 137: Print failure to journal

Error 138: PRINTER FAILED WHILE PRINTING TO THE RECEIPT PRINTER

fix the problem:

1. Verify that there is paper in the printer. Replenish paper as needed.
2. Release the printer locking screw. Open the printer support bracket and verify that there are no jams in the printer or the paper path. Verify the blue lever on the left side of the printer is in the print position.
3. Inspect the cable supplying DC power from the power supply to the printer. Make sure it is connected to CN1 on the printer control board. Verify that all DC voltages applied to the printer control board are correct. If the votages are not within tolerance, the cable or power supply may be defective.
4. Inspect the printer data cable for damage. Make sure it is seated at the main board and at the printer control board at CN6.
5. Inspect the cables connected to CN2, CN3, CN4, and CN5. Make sure they are seated at the printer control board and not damaged.
6. If the voltages are correct and the cables are undamaged and correctly connected, causes of this error may be a defective cable, a defective printer, printer control board, or main board.

Error 138: Print failure to receipt

Error 139: PRINTER CONTROLLER NOT RESPONDING TO COMMANDS

Error 14: BGD DONE BAD

Cause:

1. EOT was received as first character after the first request message was sent.
2. If EOT is received on subsequent attempts, then Error Code 1 is reported.

Result:

No money is dispensed, screen and receipt display system unavailable.

Error 140: TIME-OUT WAITING FOR PRINTER TO BE READY

Error 141: PAPER JAM REPORTED BY THE CONTROLLER DURING STATUS CHECK

fix the problem:

1. Release the printer locking screw. Open the printer support bracket and verify that there are no jams in the printer or the paper path. Verify the blue lever on the left side of the printer is in the print position.
2. Inspect the cable supplying DC power from the power supply to the printer. Make sure it is connected to CN1 on the printer control board. Verify that all

DC voltages applied to the printer control board are correct. If the votages are not within tolerance, the cable or power supply may be defective.

3. Inspect the printer data cable for damage. Make sure it is seated at the main board and at the printer control board at CN6.
4. Inspect the cables connected to CN2, CN3, CN4, and CN5. Make sure they are seated at the printer control board and not damaged.
5. If the voltages are correct and the cables are undamaged and correctly connected, causes of this error may be a defective cable, a defective printer, printer control board, or main board.

3.9 Installing the ATM Machine

ATM installation

In order to install the ATM machine, it is advised to choose the right location for it. Security concerns suggest that place of ATM installation should be away from any door, large window, or external access point. Ideally, the machine should be installed as close to a back wall as possible. For visibility reasons, it's often best to position the ATM near the front where your customers can easily locate it.

3.9.1 Tools/Equipment Required

- Heavy-duty electric drill (rotary hammer drill)
- 1/2" (13mm) diameter carbide-tipped masonry drill bit
- Four 1/2" x 4 1/2" (or optionally 1/2" x 5" or 1/2 x 7") wedge anchor bolts
- Four 1/2" nuts
- Four 1/2" flat washers
- 3/4" Wrench (Recommended: Box End Ratcheting Wrench)
- Dolly (hand truck)
- Hammer
- Sharpie or felt-tip marker
- Flashlight
- Eye protection

- Broom (or vacuum, blower, etc.)
- Tile drill bit (optional)
- 1/4" (6 mm) diameter carbide-tipped masonry drill bit (optional)
- Extra ½" wedge anchors (optional)
- Extra drill bits

3.9.2 Installation Considerations

Site Wiring

- An obvious and essential first step is making sure that there is a standard electrical outlet and proper network communication lines available nearby the spot you want to install the ATM at. Although it is not entirely necessary, it is best if AC power feed will be a dedicated electrical outlet, to which no other electrical devices are connected.
- If you are using a dedicated phone line to connect your ATM, ensure that it will be a dedicated line, not regularly used by other connections which will cause interruption.
- If you are using a TCP/IP (Internet/Ethernet) connection, ensure that the connection is communicating first by connecting it to a computer. Make sure that there are no security firewalls that would prevent communication with the Host/Processor.

Clearance and Space Requirements

- Once you have a spot picked out it is important to make sure that there is enough space around the ATM to ensure it can be easily used, loaded, and operated.
- Make sure that the fascia (the plastic door used to cover the lower cabinet and lock) and the vault door can swing open far enough so that the cash can be loaded in the cabinet.

- It is best if the door can swing open 90 degrees.
- If you're placing your ATM next to movable objects such as vending machines for example, those machines can be butted up against the ATM, then moved slightly when the ATM door needs to be opened, and moved back when it's closed.

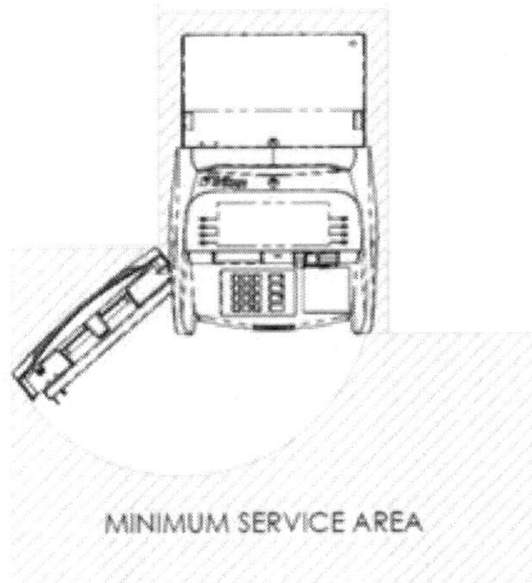

Behind the ATM you will need about an inch of space to get the power cord and wires out.

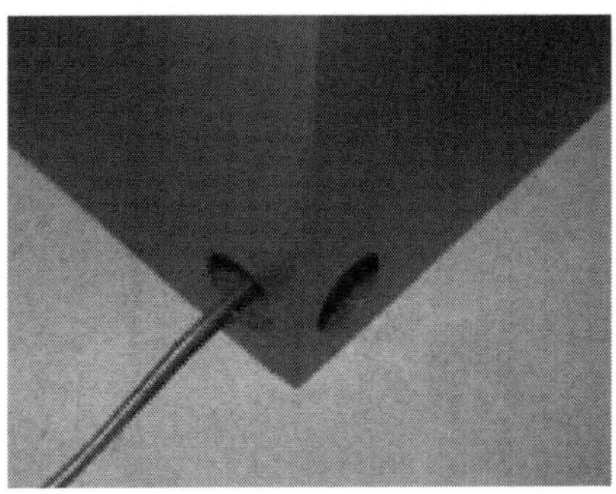

3.9.3 Installation Process

Marking the Holes

Once the site is ready, use the dolly to move the ATM into the desired position and to make sure that the ATM is in the exact position. Now, open the cabinet vault door. In the base of the ATM safe there are 4 pre-drilled holes in each corner that are used for mounting. Use a Sharpie marker or any felt-tip pen, and trace these holes onto the floor. Mark the entire hole onto the floor, these marks will serve as guides for the anchor bolt holes that will be drilled in the next step. Once you've marked on the floor where all four holes are to be drilled, move the ATM away from it's spot so it is no longer on top of the four holes. Now on the floor you can see the "bolt pattern".

Drilling the Holes

ATM can be installed with minimum number of bolts. With one bolt the ATM can move if leaned against or pushed, so it is better to apply two bolts so as to make the machine unmovable. Sometimes one bolt will be anchored in the floor a lot more strongly than another bolt, so using fewer bolts lowers the strength. For security purposes, use four bolts at four corners.

Concrete/Tile Floors:

There are certain instructions which needs to be followed while attach the ATM to concrete or tile floors. If you need to attach the ATM to hardwood you will need to follow a different process using large screws.

Drilling Through Tile

If you have a tile floor, you must not attempt to drill through the tile using the concrete drill bit. If you do this you will most likely crack the tile. You will need to use a special 1/2" carbide tipped glass & tile drill bit

Drilling Pilot Holes

Sometimes pre-drilling the holes can help make the main drilling easier. Although it is not completely necessary, if you already have the 1/4" drill bit, it might be worth it. To get started, center-punch each mark to help align the drill bit. Use a rotary hammer drill with a 1/4" (6 mm) diameter carbide-tipped masonry bit to drill four pilot holes at the drilling points that were marked in the previous step. Drill the pilot holes approximately 1/2" (12 mm) deep into the floor. These holes will help guide the masonry bit that will be used to drill the anchor-bolt holes in the next step.

Drilling the Anchor-Bolt Holes

When drilling be sure to wear safety glasses! Pieces of concrete can go flying and cause injuries. It is also recommended to wear earplugs as the drilling can be very loud.

Use a rotary hammer drill with a 1/2" (13mm) diameter carbide-tipped masonry bit to drill four holes at least 2-3/4" [70mm] deep into the floor. Be sure to take into account the depth of any floor covering, such as tile or vinyl when gauging the depth of the anchor holes. Make sure the holes are drilled at least 2- 3/4" (for a 4" long anchor bolt, deeper for longer anchors) into the concrete floor.

Hammer in Anchor Bolts

After drilling the holes, the floor is going to be covered in dust and debris from the concrete. Make sure that the location you plan to mount the ATM at is free of all debris that might cause the cabinet to not be level before continuing. Use a broom, vacuum or blower to remove any dirt or dust in the area. Be careful not to inadvertently sweep too much debris into the mounting holes as this can cause problems.

Move the ATM into position for mounting by aligning the cabinet base holes over the four anchor bolt holes drilled in the previous procedure. You will likely need a flashlight to be able to clearly see if the holes are aligned or not. Once aligned, place the four ½" anchor bolts through the cabinet base and into the mounting holes.

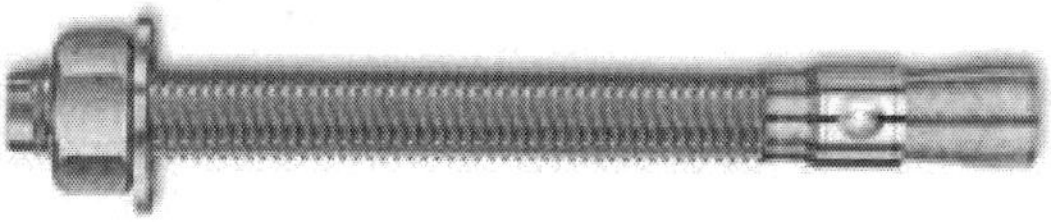

Fig Bolt

On each bolt go ahead and place a flat washer followed by a ½" nut before hammering. Sometimes the tips of the anchor bolts can be stripped or damaged during the hammering process making it impossible to fasten the nut to the bolt, therefore it is better to do this step before hammering. Once the washers and nuts are on, use a hammer to pound the bolts firmly into the floor.

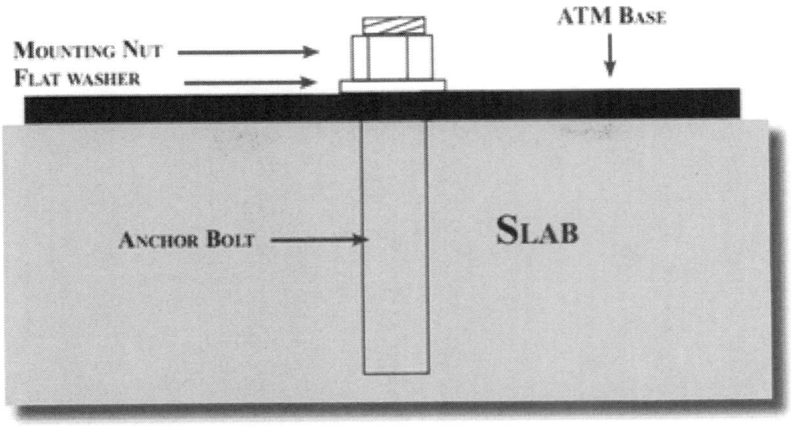

Tightening the Nuts

Although you can usually use any type ¾" wrench for this step, we have found that it works best to use a Box End Ratcheting Wrench. Wrenches with sockets are often not deep enough when there is excess bolt remaining and the lack of space inside the ATM cabinet often makes it difficult to use a standard or crescent wrench.

If you're using a torque wrench, it is recommended the bolts be tightened to 60 foot-pounds to establish the maximum pull-out strength of the anchors. If a torque wrench is not available, do not over tighten. If you over tighten the nuts it will often cause the anchor bolts to spin in the hole.

When you've finished installing the ATM make sure that the ATM cannot be moved by giving it a good push. Now that the physical installation of your ATM is completed, the hard part is over. The next step is to connect all the electrical and network cables and program the machine.

Chapter 4
Security aspect of Automated Teller Machine (ATM)

4.1 Introduction

Security, as it relates to ATMs, has several dimensions. ATMs also provide a practical demonstration of a number of security systems and concepts operating together and how various security concerns are dealt with. As the number of ATM units increase, the machines are prone to hacker attacks, fraud, robberies and security breaches. In the past, the ATM machines' main purpose was to deliver cash in the form of bank notes and to debit a corresponding bank account. However, ATM machines are becoming more complicated, and they serve numerous functions, thus becoming a high priority target to robbers and hackers.

Fig Robbery at ATM

Modern ATM machines are implemented with high-security protection measures. They work under complex systems and networks to perform transactions. The data processed by ATMs are usually encrypted, but hackers can employ discreet hacking devices to hack accounts and withdraw the account's balance. As an alternative, unskilled robbers threaten bank patrons with a weapon to loot their withdrawn money or account.

ATM machines are now turning increasingly vulnerable to all kinds of malware strikes. This happens in addition to the other frauds committed at ATM counters, including card skimming, using hidden cameras etc. Cybercriminals manage to infect ATM software with malware using different techniques and making use of different kinds of vulnerabilities. This has been happening worldwide for the last few years and now it's showing an increasing trend. Thus, securing ATM software becomes as important as securing the machine itself.

4.2 Physical

Early ATM security focused on making the ATMs invulnerable to physical attack; they were effectively safes with dispenser mechanisms. A number of attacks on ATMs resulted, with thieves attempting to steal entire ATMs by ram-raiding. Since late 1990s, criminal groups operating in Japan improved ram-raiding by stealing and using a truck loaded with a heavy construction machinery to effectively demolish or uproot an entire ATM and any housing to steal its cash.

Another attack method, plofkraak, is to seal all openings of the ATM with silicone and fill the vault with a combustible gas or to place an explosive inside, attached, or near the ATM. This gas or explosive is ignited and the vault is opened or distorted by the force of the resulting explosion and the criminals can break in.

Modern ATM physical security, per other modern money-handling security, concentrates on denying the use of the money inside the machine to a thief, by means of techniques such as dye markers and smoke canisters.

Fig ATM thief

A common method is to simply rob the staff filling the machine with money. To avoid this, the schedule for filling them is kept secret, varying and random. The money is often kept in cassettes, which will dye the money if incorrectly opened.

4.2.1 Transactional secrecy and integrity

The security of ATM transactions relies mostly on the integrity of the secure cryptoprocessor: the ATM often uses commodity components that are not considered to be "trusted systems".

Encryption of personal information, required by law in many jurisdictions, is used to prevent fraud. Sensitive data in ATM transactions are usually encrypted with DES, but transaction processors now usually require the use of Triple DES. Remote Key Loading techniques may be used to ensure the secrecy of the initialization of the encryption keys in the ATM. Message Authentication Code (MAC) or Partial MAC may also be used to ensure messages have not been tampered with while in transit between the ATM and the financial network.

4.2.2 Customer identity integrity

There have also been a number of incidents of fraud by Man-in-the-middle attacks, where criminals have attached fake keypads or card readers to existing machines. These have then been used to record customers' PINs and bank card information in order to gain unauthorized access to their accounts. Various ATM manufacturers have put in place countermeasures to protect the equipment they manufacture from these threats.

Alternate methods to verify cardholder identities have been tested and deployed in some countries, such as finger and palm vein patterns, iris, and facial recognition technologies. However, recently, cheaper mass production equipment has been developed and being installed in machines globally that detect the presence of foreign objects on the front of ATMs, current tests have shown 99% detection success for all types of skimming devices.

4.2.3 Device operation integrity

Openings on the customer-side of ATMs are often covered by mechanical shutters to prevent tampering with the mechanisms when they are not in use. Alarm sensors are placed inside the ATM and in ATM servicing areas to alert their operators when doors have been opened by unauthorized personnel.

Rules are usually set by the government or ATM operating body that dictate what happens when integrity systems fail. Depending on the jurisdiction, a bank may or may not be liable when an attempt is made to dispense a customer's money from an ATM and the money either gets outside of the ATM's vault, or was exposed in a non-secure fashion, or they are unable to determine the state of the money after a failed transaction. Bank customers often complain that banks have made it difficult to recover money lost in this way, but this is often complicated by the bank's own internal policies regarding suspicious activities typical of the criminal element.

4.2.4 Customer security

Thieves and criminals are always coming up with new ways to gain access to your most sensitive information. To authenticate the identity of e-banking customers, bank shall employ techniques appropriate to the risks associated with its products and services. The implementation of appropriate authentication methodologies should start with a risk assessment process. The risk should be evaluated based on the type of customer; the customer transactional capabilities (e.g., bill payment, fund transfer, inquiry); the sensitivity of customer information and transaction being communicated to both the Bank and the customer; the ease of using the communication method; and the volume of transactions.

Account fraud and identity theft are frequently the result of single-factor (e.g., ID/password) authentication exploitation. Where risk assessments indicate that the use of single-factor authentication is inadequate, Security Bank shall implement multifactor authentication (e.g., ATM card and PIN), layered security, or other controls reasonably calculated to mitigate those risks.

Bank's authentication process shall be consistent with and support overall security and risk management programs. An effective authentication process shall have customer acceptance, reliable performance, scalability to accommodate growth, and interoperability with existing systems and future plans as well as appropriate policies, procedures, and controls.

In some countries, multiple security cameras and security guards are a common feature. In the United States, The NY State Comptroller's Office has criticized the NY State Department of Banking for not following through on safety inspections of ATMs in high crime areas. Critics of ATM operators assert that the issue of customer security appears to have been abandoned by the banking industry; it has been suggested that efforts are now more concentrated on deterrent legislation than on solving the problem of forced withdrawals.

4.3 ATM Security Measures

4.3.1 Physical security measures

Physical ATM security is of importance in cases where cipher keys reside in terminals. In the absence of physical security, an abuser may probe for a key or substitute its value. To avoid such abuses, banks should preserve the integrity of non-secret parameters as well as the confidentiality of secret parameters. ATM security should focus on protecting ATMs from physical attacks. Modern ATM security concentrates on denying access to money inside the machine to a thief, by means of techniques such as dye-markers and smoke canisters.

Other methods are:

1. Perimeter surveillance
2. Access control
3. Intrusion detection
4. Tested and approved ATM enclosures
5. Security guards
6. Central monitoring station

PCI perspective of physical ATM security: According to PCI PTS, the following factors must be considered for ATM security:

- ❖ Installation of a physical shielding barrier. This means that you need a tamper-proof casing not only for the ATM but also the PIN pad.

- ❖ Implementation of dual security mechanisms
- ❖ Proper environmental and operational conditions
- ❖ Controlled and monitored physical access
- ❖ Quick incident response mechanism

4.3.2 Logical security measures

Network plays a key role in the functioning of ATMs since a customer swipes his card, enters the PIN and details are then sent to the RDBMS for validation. Attackers usually intercept this information to perform logical frauds. The following logical security measures can help prevent such incidents:

1. Firewall (s)
2. Effective tracking and monitoring system
3. Encryption technologies
4. Logical access control
5. Fraud detection system
6. Protection of communication

PCI perspective for logical ATM security: The following aspects come into play when you look at logical ATM security from a PCI standpoint.

- ❖ Acquirers, processors and other third-parties that have access to (store, process and/or transmit cardholder data) must comply with PCI DSS.

1. All must currently comply with the PCI DSS requirements of "Do not store magnetic stripe data, pins or pin blocks".
2. Check ATM audit logs to ensure only required (or allowed) data are stored.
3. Caution: If your organization or parties sponsored by your organization is compromised, you may be subject to the fines under PCI DSS, civil and statutory damages.

- ❖ Track all parties with access to cardholder data (excluding telephone and cable companies)
- ❖ Maintain inventory of all devices and applications

ATM manufactures must ensure that their devices comply with PCI PIN and PTS requirements. In addition, the PA DSS may eventually impact all software installed on ATM devices as well.

On the PCI PIN front, it's critical to implement, maintain and secure the PIN. The key items to note on this front are:

- Dual control and split knowledge: Prevents fraud and identify theft
- Periodic attestation
- Use of devices that are:

1. TDES/AES encrypted
2. Tamper Resistant Security Modules (TRSM) to prevent compromise of cryptographic security parameters

Chapter 5
Working of Automated Teller Machine (ATM)

5.1 Introduction

An ATM (Automated Teller Machine) is a computerized system individuals can use to deposit money into their banking accounts, withdraw money from their accounts, check balances and transfer money from one account to the other. ATMs can either be part of a specific bank (a Suntrust ATM, for example), or an independent entity that has contracts with several banks and credit card companies (a Presto machine, for instance). The former are usually found in stand alone booths or on the outsides of a bank, while the latter are usually found in other public places, such as grocery stores and malls.

Fig ATM in public places

Before proceeding to know how ATM machine works, let us know what is ATM and why do we need these machines in our day to day life. ATM stands for Automated teller machine. As its name suggests, it is an automated machine which can process your requests of withdrawing money, transferring money to other accounts, balance enquiry etc. The invention of these machines although took place in 1932, Tokyo was the first city to implement it successfully in the year 1962. After that rapid developments took place and now we are reaping its benefits. Automated teller machine is a machine which is used to withdraw money, transfer money and to know the balance, mini statement record of transactions, and even to recharge our mobile phones. So we need these automated teller machines to accomplish our day to day financial transactions without standing in the long queues of bank. So automated teller machines have become part and parcel of our life.

5.2 Components of ATM Machine

ATM machine do have various major components which are responsible for the functioning of the Automated teller machine. Let us see the major parts of ATM in detail.

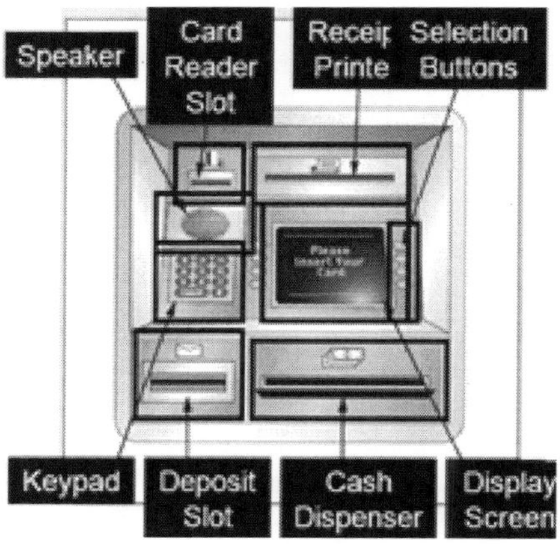

Fig ATM parts

Card reader

Card reader is one of the most important and the most basic component of automated teller machine. If there will be no card reader, then ATM cards or debit cards cannot be read and so no transactions can be done trough this machine. ATM or debit card is required to swipe through this card reader slot. Then it captures your account information through the magnetic stripe of the ATM or debit card. After retrieving the correct information of your account, card reader send this information to the host processor for further process.

Host processor

Host processor is the heart of ATM machine. It contacts the bank server and retrieves your account details such as available balance, previous transactions, PIN etc. It verifies your PIN number and if it matches with the provided PIN number from the bank server, you are allowed to make the transaction. So in order to take your PIN and other requests, it does have a keypad or touch pad.

Keypad or touch pad

Keypad or its latest counter part touch pad is responsible for taking input from the user of ATM machine. On this pad, numbers and few other dedicated keys such as Clear, OK etc., will be placed. You have to enter your PIN and the amount, account number in case of transferring the money using this touch pad or keypad. Key pad transfers your inputted data to host controller to process the request.

Speaker

Speaker is placed in ATM machine which is responsible for audio alerts. When you press any key, this speaker confirm the pressing of key by producing a beep sound alert. It assures you that the key pad is working. So speaker is also very important part of ATM machine.

LCD or CRT screen

Another major integral part of the ATM machine is its Cathode Ray Tube (CRT) or Liquid Crystal Diode (LCD) screen. CRT or LCD screen is responsible for visual display of your transactions. It is responsible for displaying your name, entered amount, account number etc. Without CRT or LCD display screen, the transactions from ATM machine would have become more time consuming and difficult.

Receipt printer

Receipt printer is responsible for the printing of your transaction's receipt. It will print whether the transaction is successful or not. It does contain the crucial information in the form of fixed codes by which we can assess the transaction status and its reason of failure. SO receipt printer is a vital part of automated teller machine.

Cash dispenser

Last but not least, and the most important part of automated teller machine is its cash dispenser. What is the use of an ATM machine if you are not able to take the money out of it. Yes! the cash dispenser is responsible for giving out the correct amount of cash you have requested for. So after getting instruction from host controller to pay the money, cash dispenser will count the money and dispenses the money out of it.

5.3 Principle of ATM machine

ATM is a packet technology that directs traffic using a label contained in the packet header. Unlike other packet technologies, ATM uses short, fixed-length packets called cells. Each cell is 53 bytes long: 48 bytes for the information field and 5 bytes for the preceding header. The header field contains information about the virtual channel and virtual path in use, payload type and cell loss priority.

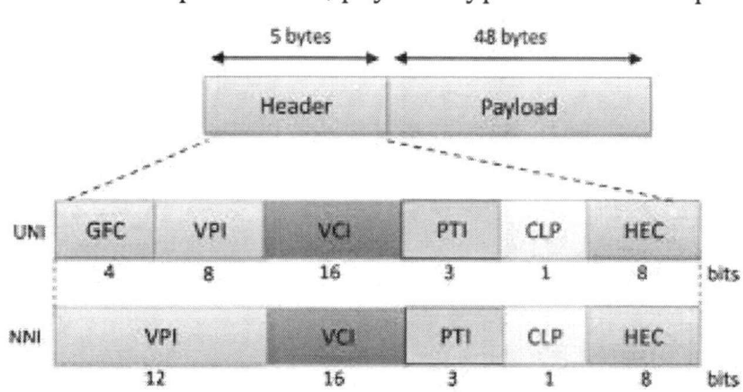

Fig ATM cell structure

The basic idea of ATM is to segment data in small cells and then transfer them by the use of cell-switching. Such cells have a uniform layout and a fixed size of 53 bytes, which greatly simplifies switching. Being more complex, packet-switching is not nearly fast enough to be of use for isochronous data (i.e. real-time video and sound). Cell-switching gives maximum utilization of the physical resources.

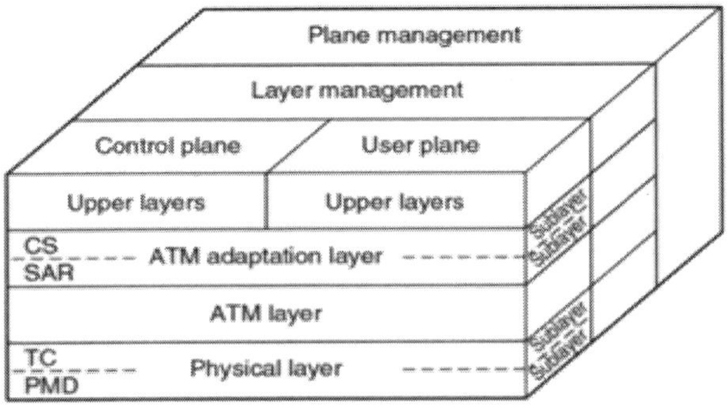

Fig Layers of data segmentation

The basic principles of this technology were first formulated by AT&T and the French Telecompany in the first half of the 80's. Researchers found that if data was segmented in small, fixed size cells, switching could be done with simple, specially designed ICs. With sufficient intelligence to handle routing information in each cell, such ICs could be used to build very fast switch-matrixes. By connecting several switch-matrixes, highly efficient networks with small and predictable transmission delays can be built. The fact that ATM can be used efficiently in both WANs and LANs shows how powerful ATM is.

When two participants wants to talk to each other, they first have to set up a connection between them. This connection remains open during the whole session, and when the session is over the connection is closed. To the user it appears that the connection is permanent once established, but it's actually not. Cells are switched through the network so quickly that it appears as if there is a direct connection. To be able to transmit a message to a host by ATM, a connection needs first to be made. ATM is thus connection-oriented.

Cell

An ATM cell has 5 bytes of header (administrative information) and 48 bytes of payload (data), counting a total of 53 bytes. The choice of the cell size caused much controversy during CCITT standardization. The US computer branch wanted 64 bytes on payload, because they were considering the bandwidth utilization for data networks and efficient memory transfer (length of payload should be a power of 2 or at least a multiple of 4). 64 bytes fit both requirements.

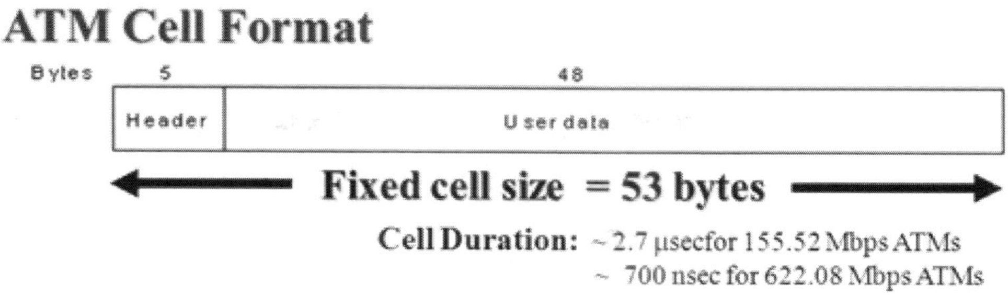

Fig ATM cell format

Representatives from across the globe telephony branch was taking voice applications into consideration and wanted smaller cells. At cell sizes greater than 151, there is a talker echo problem. Cell sizes between 32 and 152 result in listener echo. Cell sizes less than or equal to 32 overcome both problems, under ideal conditions. Europeans telecommunication companies had no desire to invest money in echo cancelling equipment, and thus went for 32 bytes of payload.

However, because of big distances the US telecommunication companies had already installed echo cancellers across the country, and didn't consider this to be a problem. After a lot of discussion 48 bytes of payload was agreed on. As far as the header goes, 10% of payload was perceived as an upper bound on the acceptable overhead, so 5 bytes was chosen. Although the header of 5 bytes is much for a 53-byte cell, the flexibility is worth the cost.

The header of 5 bytes contains 24 bits of VCI label, 8 bits of control field, and 8 bits of checksum.

GFC: Generic Flow Control.

VPI: Virtual Path Identifier.

VCI: Virtual Channel Identifier.

PT: Payload Type.

Indicates what the cell contains. Also used for flow control.

CLP: Cell Loss Priority.

Indicates whether the cell should have high or low priority in the case of overload.

HEC: Header Error Correction.

For error detection and correction. Only 1 bit errors can be corrected.

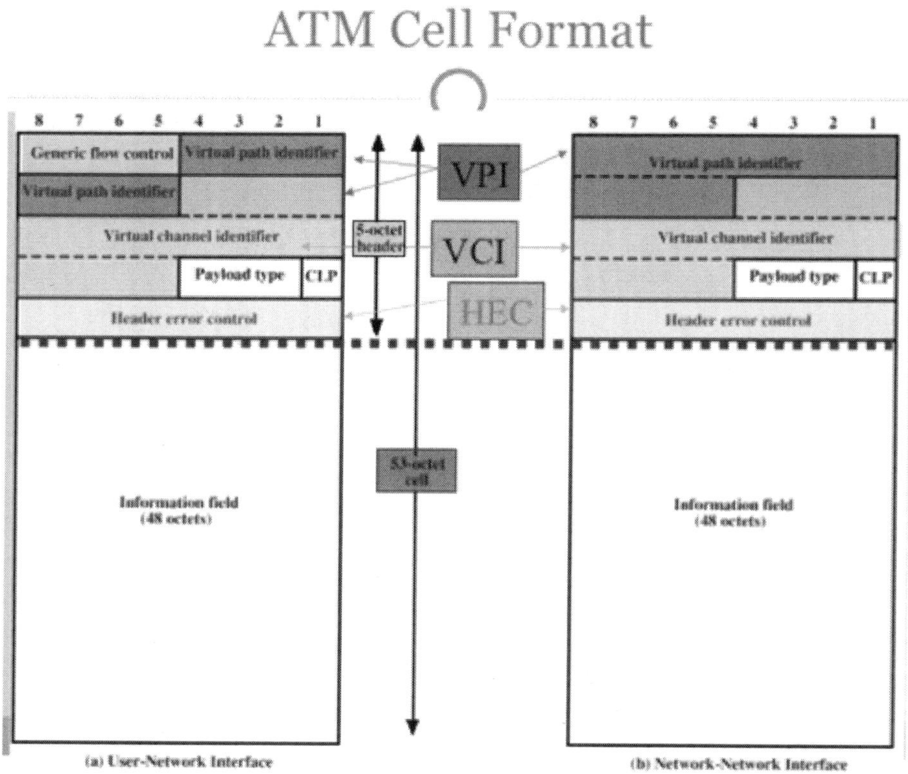

Fig Cell format

The 48 bytes of payload may optionally contain a 4-byte ATM adoption layer and 44 bytes of actual data, or all 48 bytes may be data. This depends on a bit in the control field of the header. This enables fragmentation and reassemble of cells into larger packets at the source and destination respectively. The control field may also contain a bit to specify whether there is a flow control cell or an ordinary cell, an advisory bit to indicate whether a cell is droppable in the face of congestion in the network or not.

It is interesting to note that there is no sync character in the ATM cell. In order to stay in sync with the data stream, the receiving host calculates the checksum for incoming cells on the fly. If connection has been lost, the host starts computing the checksum for everything received from the wire. When the checksum has been computed correctly for 16 subsequent cells it is reasonable to assume that the host is in sync.

5.4 Working Procedure

In order to access various accounts, ATM computers are connected to complex interbank networks, which allows customers from many different banks to access their accounts through the machine. There are several of these interbank networks;

among them are Cirrus and LINK. In order to provide access to the customer, ATMs rely on authorization from the financial institutions. To cover the costs of operation and the fees that involve belonging and accessing the networks, most independent ATMs charge a fee for each transaction.

Fig ATM charges

These days every ATM contains a computer with special software integrated into it, which regulates the dispensing of cash while also managing queries and other transactions. This is because, in reality, ATMs are no longer just a tool used to withdraw and deposit cash. They can now also be used to view account transactions and carry out many other banking procedures.

Fig Cash dispensing

There are many types of cash dispenser machines and most of them feature a screen, a card slot, a cash dispensing slot, a keypad, and some keys at the side of the screen. Many newer-model ATMs have cameras as well.

As soon as you start to use an ATM, you set off a sort of technological chain reaction. Inserting your card into the proper slot is the first step. As soon as you do this, the card will be read by a special device and a connection with the bank will be automatically established. Next, after you have completed the security-oriented step of entering your personal ID number or PIN, you will be given access to access a variety of options.

Fig Entering PIN

For example, suppose you are a customer who wants to withdraw some cash. This is of course the main purpose for which ATMs are used, as well as the one that primarily justifies their existence. This is especially true because banks have by now created extensive networks that let you withdraw cash from ATMs operated by other banks (although you will often have to pay a fee in order to do this).

If you select the cash withdrawal option, the ATM will determine how much money you want by asking you to enter an exact amount. The machine will then count and dispense your cash right away. This process is able to take place thanks to a precise combination of more than 400 components, which must work together like clockwork to perform the series of seven steps required in order to provide your cash:

- ❖ Customers use the keypad to enter the amount of cash they want (and in some cases they can even select the denominations and combinations of bills they would like to receive).
- ❖ Next, the ATM checks the customer's balance and then authorizes the withdrawal transaction as long as there are sufficient funds available in the account.
- ❖ Inside the ATM, cash is stored in a series of trays, which allow the bills to be made available to the dispensing mechanism.
- ❖ A complex gear system allows the bills to be taken from the appropriate tray one at a time.
- ❖ An optical sensor then checks the amount while also confirming that none of the bills are counterfeit.
- ❖ A small slot or door then opens and the cash requested is delivered to the customer.

- Next, the ATM asks whether or not the customer wants a printed receipt for the transaction, then after responding appropriately based upon the customer's response, it closes the process and disconnects the session.

5.5 Determination of correct denominations

ATM is an electro-mechanical device, with a switch that knows how much cash is available in an ATM and in what denomination. In a cash withdrawal transaction, once your PIN is verified and there is sufficient balance in your account, the switch sends an instruction to the ATM to dispense the amount requested. You would be receiving a mix of Rs2,000, Rs500, Rs100 and sometimes even Rs200 from ATMs nowadays.

Fig Cash determination

Have you ever wondered how the ATMs know which denomination of notes to dispense and how many of each denomination? Most of the switches follow the principle of 'last note to be dispensed in the lowest denomination available in the ATM' at the time of withdrawal'. For example, if you want to withdraw Rs1,000 from an ATM and the ATM has Rs500 and Rs100 denomination available, it will dispense one note of Rs500 and five notes of Rs100. If instead of Rs500 and Rs100, it has Rs200 and Rs100, the ATM will dispense four notes of Rs.200 and two notes of Rs100.

If the ATM has three denominations, say Rs500, Rs200 and Rs100 at the time of withdrawal, it will dispense one note of Rs500, two notes of Rs200 and one note of Rs100.

The currency notes inside the ATM are stacked in boxes called 'cassettes'. Each cassette is loaded with one denomination. Most of the ATMs in the country have four such cassettes, while some of the older ATMs may have two cassettes. To enable the ATM to understand what denomination is available and in which cassette, the cassettes need 'calibration', based on the length, breadth, thickness of the notes they are supposed to hold. This is done by trained technicians either at the ATM factory or at the ATM site.

The cassettes in the ATM are loaded with currency notes on a regular basis by the bank staff or their authorized agencies. Banks are mandated to issue clean and crisp (ATM-fit) notes and ensure the notes are genuine. The cash is carried from the currency vaults of banks to the ATMs, in security vans with armed escorts.

The amount of cash that is loaded in the ATM depends on various factors such as average number of transactions on the ATM in the past, average value of withdrawal, beginning of the month (salary period), festivals, long weekends, distance of the ATM from the cash vault etc.

When you withdraw cash from an ATM, you may hear a lot of noise emanating from inside the machine. This is the noise generated when the dispenser arm of the ATM picks out the number of notes required from the cassettes. The notes are picked up one by one from the cassette and put on the dispenser belts which move and take the notes up to the dispensing outlet of the ATM.

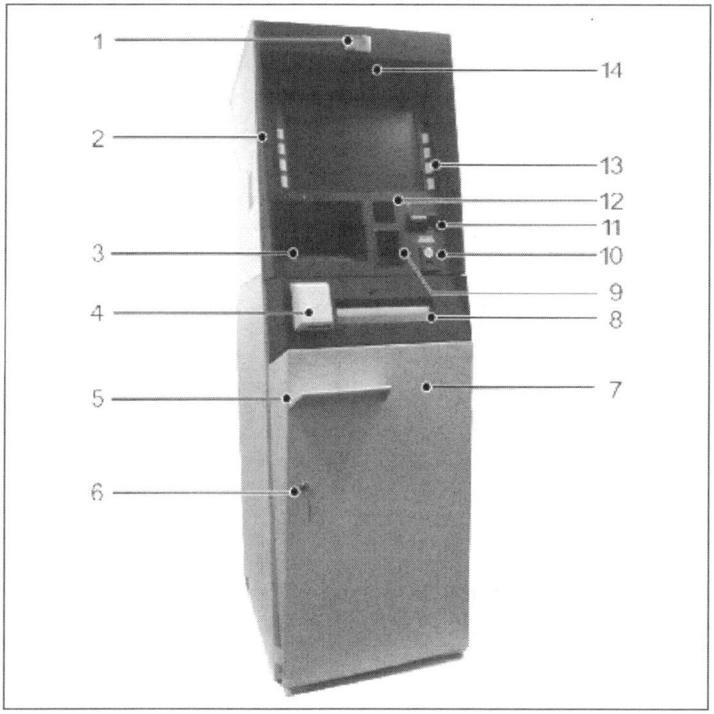

1 Mirror
2 Customer panel
3 EPP keypad
4 Barcode reader
5 Barcode table
6 Front door lock
7 Front door
8 Cash output
9 Finger print biometric reader
10 Headphone jack
11 Card input with optical indicator
12 Portrait camera
13 Soft keys
14 Cash slot camera

Fig Front View of ATM

While ATMs dispense cash accurately, in the rarest of rare cases, there could be an error owing to which all notes may not get dispensed to you. Although it may seem an overkill, it is advisable to count the cash received from an ATM. Take care to do it quickly else the customer standing behind you may get irate.

Chapter 6
Different Services Offered by ATM

6.1 Introduction

An automated teller machine (ATM) is an electronic banking outlet, which allows customers to complete basic transactions without the aid of a branch representative or teller. Anyone with a credit card or debit card can access most ATMs.

There are two primary types of ATMs. Basic units allow customers to withdraw cash and receive reports of their account balances only. The more complex machines accept deposits, facilitate line of credit payments and report account information. To access the advanced features of the complex units, a user must be an account holder at the bank that operates the machine.

Analysts anticipate ATMs will become even more popular and forecast an increase in the number of ATM withdrawals. ATMs of the future are likely to be full-service terminals instead of or in addition to traditional bank tellers.

In many cases, banks and credit unions own ATMs. However, individuals and businesses may also buy or lease ATMs, on their own or through an ATM franchise. When individuals or small businesses such as restaurants or gas stations own ATMs, the profit model is based on charging fees to the machine's users. Banks also own ATMs with this intent, but in addition, the convenience of an ATM is a service banks use to attract clients. ATMs also take some of the customer service burden off bank tellers, saving banks money in payroll costs.

These days ATMs provide variety of services in addition to cash withdrawal, the service that they were invented for. ATMs of Indian provide following services.

- ❖ Cash Withdrawal
- ❖ Balance Enquiry
- ❖ Statement Enquiry (printing in some cases)
- ❖ PIN Change
- ❖ Cash Deposit
- ❖ Cheque Deposit
- ❖ Funds Transfer
- ❖ Credit Card Payment
- ❖ Utility Bill Payment
- ❖ Cheque Book Request
- ❖ Insurance Premium Payment
- ❖ Mobile, DTH, etc recharges and topups
- ❖ NGO donations in some cases
- ❖ Term deposit opening
- ❖ User details updation

6.2 Services

There are many useful services that can be availed from an ATM, apart from cash withdrawal and checking account balance.

6.2.1. Open or withdraw a fixed deposit

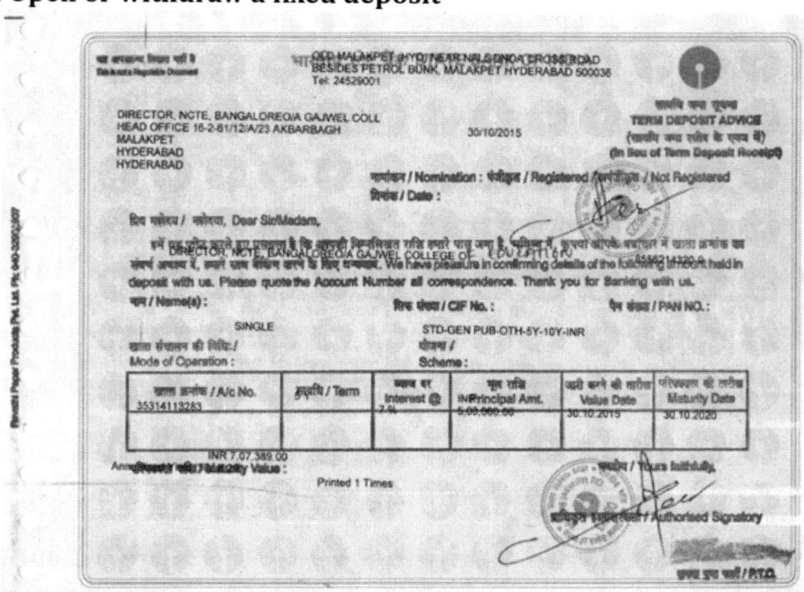
BESIDES PETROL BUNK, MALAKPET HYDERABAD 500036
Tel: 24529001
DIRECTOR, NCTE, BANGALOREGIA GAJWEL COLL
HEAD OFFICE 16-2-61/12/A/23 AKBARBAGH
MALAKPET
HYDERABAD
HYDERABAD
30/10/2015
TERM DEPOSIT ADVICE
Nomination : Registered / Not Registered
Date :
Dear Sir/Madam,
We have pleasure in confirming details of the following deposit held in deposit with us. Please quote the Account Number all correspondence. Thank you for Banking with us.
Name(s) :
CIF No. :
PAN NO. :
SINGLE
STD-GEN PUB-OTH-5Y-10Y-INR
Mode of Operation :
Scheme :

A/c No.	Term	Interest @	INR Principal Amt.	Value Date	Maturity Date
35314113283		7.%	5,00,000.00	30.10.2015	30.10.2020

INR 7,07,389.00
Maturity Value :
Printed 1 Times
Yours faithfully,
Authorised Signatory

Fig Opening Fixed deposit

You can open a fixed deposit with your bank using an ATM.

Select 'Open Fixed Deposit' on the ATM menu, select the duration, enter the amount and confirm the other necessary details.

6.2.2. Recharge your mobile

Prepaid services of most mobile operators can be recharged from an ATM.

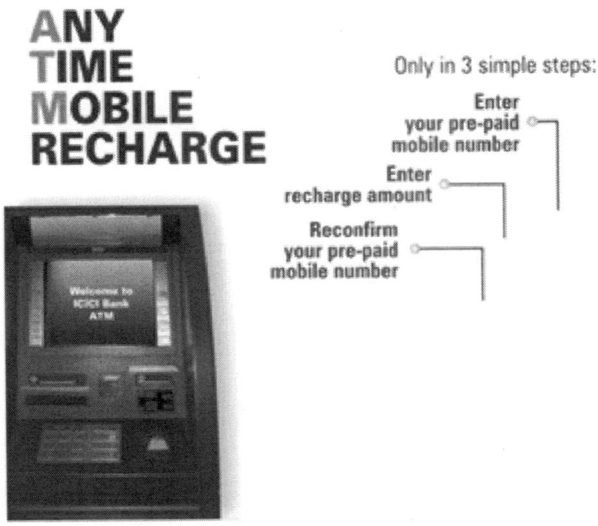

Fig Mobile recharge at ATM

Steps:

- 1. Visit any ATM Machine, insert your ATM card properly, wait for few seconds, you will be ask to Enter Your Pin number at the given box
- 2. Be sure your Enter your four (4) digits Pin number correctly and Press Proceed button.
- 3. From the ATM menu, just select Recharge Option displayed on the Screen.
- 4. Select or choose Your Preferable network, which is the network you want to recharge. For instance, GLO, MTN, ETISALAT, AIRTEL etc.
- 5. Type in the amount of airtime you want to top up on your line, then Press Proceed button.
- 6. A new pop up appears asking you to confirm your mobile phone number, please make Sure you enter correctly your mobile phone number else, you airtime goes to another person, then Press Proceed button to Finish up the Transaction.
- Finally, your mobile phone number will be credited within 30 seconds of Transaction.

Note:

- ❖ If you are using a Komerční Banka ATM and do not wish to receive a confirmation SMS, enter the code 000000000 in this section.
- ❖ Select the amount you want to recharge. You can select from the available values or choose an amount from CZK 201 - 9,999.
- ❖ If you have entered a second telephone number, the credit confirmation will be sent to the number you have entered (this could differ slightly depending on the particular ATM)

Recharging:

- ❖ Actual recharging can be done by two ways, depending on the type of ATM:
- ❖ Recharge Directly: Selected amount is added directly to the 02 prepaid telephone number you entered.
- ❖ Recharge Indirectly: ATM machine prints out a voucher with a fourteen-digit code that can be used in the same way as a regular 02 prepaid coupon: by dialing *102*XXXXXXXXXXXXXX # (X = fourteen-digit code from the voucher) by calling the Customer Care Center at 800 02 02 02 and following the voice prompts recharging via the Internet

6.2.3. Pay income tax

To make the process of tax payment more feasible, now the facility of tax payment is available in Bank ATM, which means, now you can deposit your income tax through ATM which includes advance tax, self-assessment tax as well as tax due after regular assessment. Some banks offer the convenience of paying income tax using ATMs which covers advance tax, self-assessment tax as well as tax due after regular assessment.

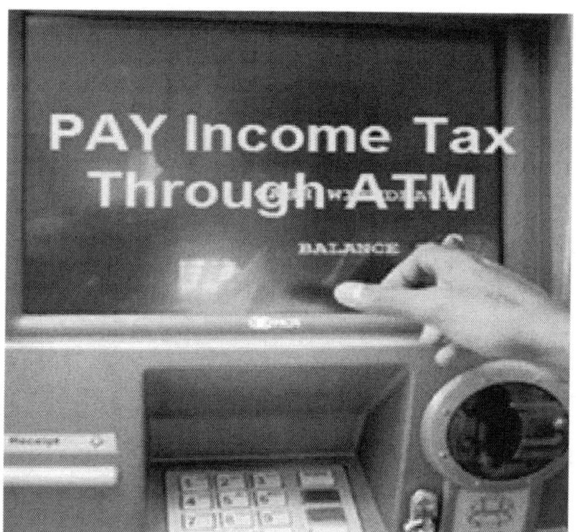

Fig Paying Tax at ATM

For this, you need to register yourself at the official website of the bank from where you wish to make the payment of tax. The bank website in turn linked to the National Securities Depositories Ltd which will help validate the permanent account number (PAN) of individuals and the Tax Deduction Account Number (TAN) provided to taxpayers.

Steps:

- After successful registration, you can visit to the Bank ATM to make your tax payment.
- To make payment using ATM, locate the income tax menu on the ATM screen.
- On selection of income tax menu, ATM screen will show you your PAN number and will ask the amount of tax that is to be paid against the PAN number.
- Enter the assessment year and the amount of tax (item wise details).
- After entering all the tax details you can click ok and the the amount of tax will be deducted from your bank account and ATM will generate a unique number (CIN) on acknowledgement receipt which can be used by the tax payer to print challan from the bank's website after 24 hours.

6.2.4. Deposit cash

When you need to deposit cash or checks, your easiest option might be to use a deposit-enabled ATM. There's no need to visit a branch or wait for the mail to get your checks to the bank. Some people don't know that depositing funds at an ATM is possible, while others are uncomfortable with the idea or they just don't know how to do it.

Machines where we can deposit cash are called as Cash Deposit Machines or just like Automated Teller Machines which helps to deposit cash directly into the machine without much hassle. This is a convenient way and saves time as of now to deposit. There is also no money that is charged and it saves you the time to deposit across the cash counter in a bank. One can deposit upto Rs 49,900/ per transaction. You can deposit cash in your account with or without ATM card.

Deposit Cash with ATM Card

Deposit cash into machine are called Cash Deposit Machines(CDM). It looks like Automated Teller Machines(ATM). So you can deposit cash directly into machines(CDM) without any problem. It is very convenient way to deposit money into your bank account as it is save your time also. This type transaction is totally free only you have to need a ATM card of that particular bank account.

Step1: Swipe your ATM card into CDM machine.

Fig Swipping of ATM in machine

Step2: Select your language

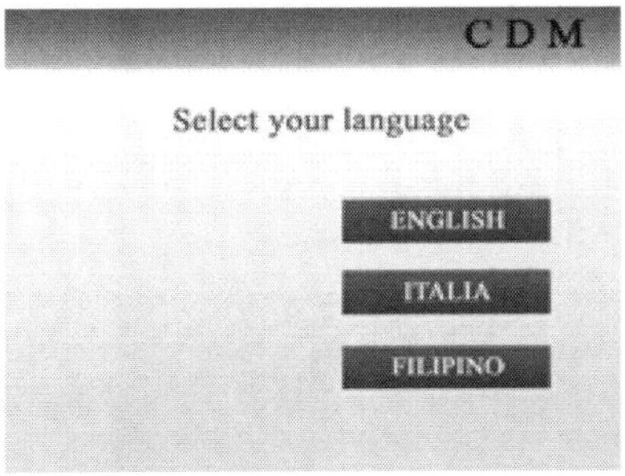

Fig Language selection

Step3: Enter your secure pin number

Fig Entering pin number

Step4: Now from the main menu Select Deposit option

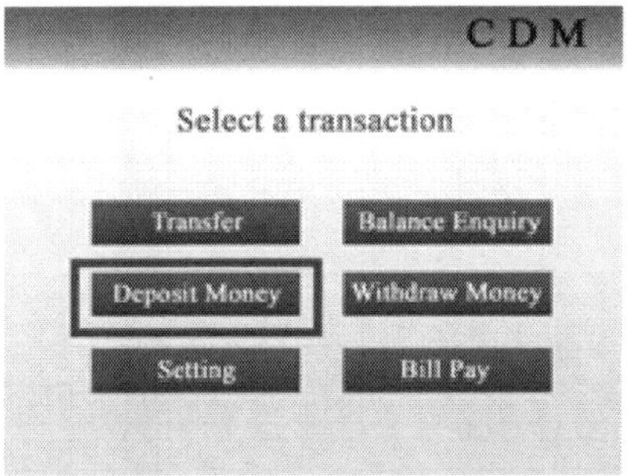

Fig Selecting Deposit option

Step5: Next you have to select cash deposit option

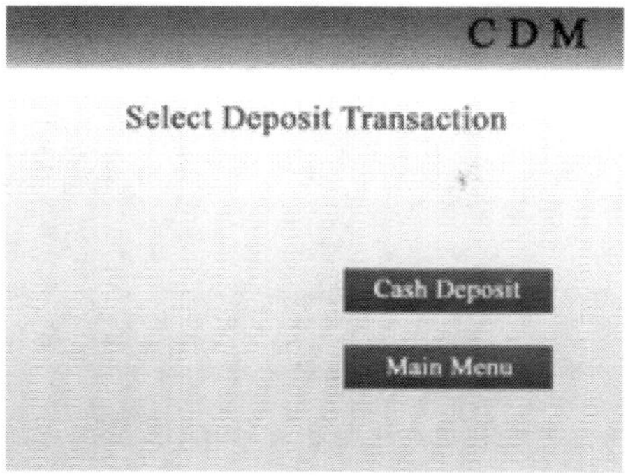

Fig Selecting Cash Deposit option

Step6: Select savings or current type of Account

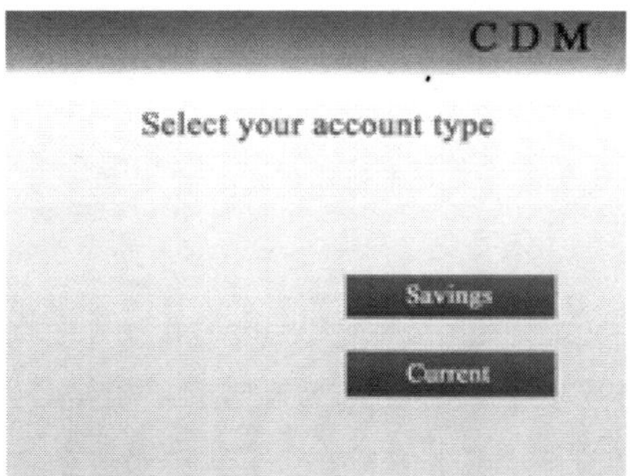

Fig Selecting Account Type option

Step7: Now press Confirm from the menu

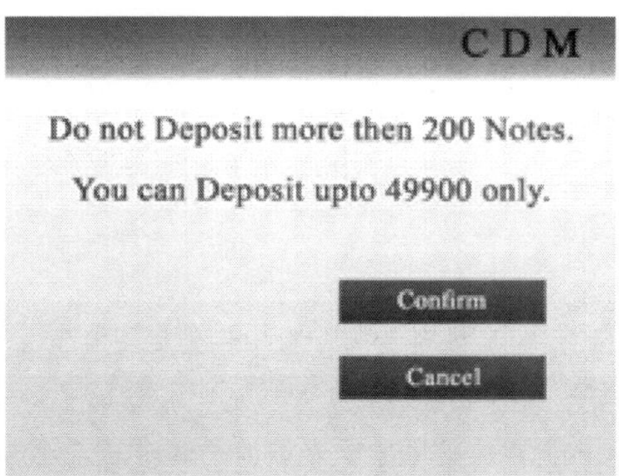

Fig Confirming menu option

Step8: Place the cash

Fig Inserting Cash in ATM

Step9: The machine may take time to validate the cash

Step10: Displays the denomination. Here you can see the summary, how many notes you enter.

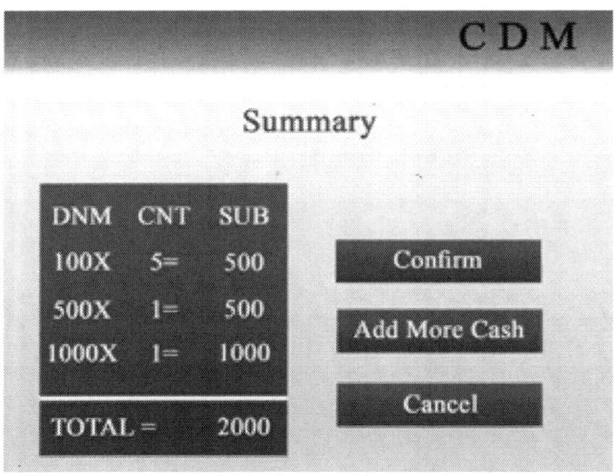

Fig Denomination summary

Step11: In Summary if you want to deposit more cash then choose " Add more Cash"

Step12: Place the cash

Step13: Click on Confirm to finish your transaction

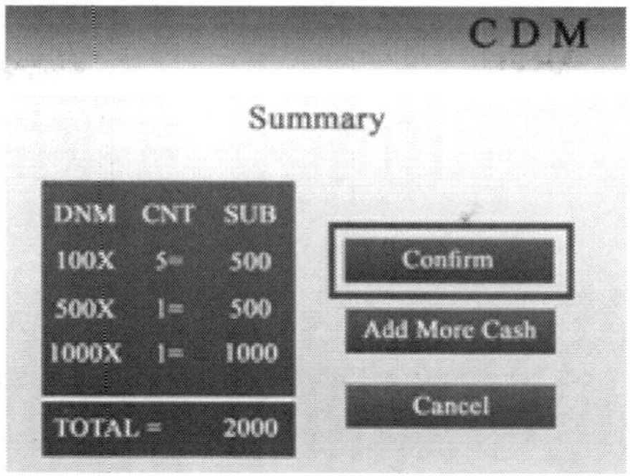

Fig Confirming Status

Deposit Cash without ATM Card

The minimum amount that can you deposit is Rs. 100 and the maximum amount is below Rs 50000. The cash deposit machine will accept all currency notes in multiples of Rs. 100 up to Rs. 2,000. You can deposit money into your account using CDM without debit card by following the steps:

Fig Currency

Step 1 Walking to the nearest cash deposit machine and check for the option 'deposit without card' on CDM and select it. After selecting the option, the machine will ask you to enter your bank account number.

Fig Checking CDM options

Step 2 Enter your bank account number; then you press the amount you wish to deposit and confirm the amount. You have to place the cash in the deposit slot and let the machine take it in because you are not using debit card here.

Fig Entering Bank Account Details in ATM

Step 3 The machine will take the money, and it will display the currency denominations received on the screen also. The depositor will get a confirmation slip for the completion of the transaction.

Fig confirmation slip of transaction

If you want to make payments to different accounts, you can do it via cash deposit machine. For this, you need to follow the same steps. But it is important to note that you need to know the account numbers you wish to transfer the money to.

Be Careful when depositing cash at ATM

It is extremely important to remember that you should exercise caution, when you deposit cash at the ATM machine. Also, make sure that the ATM is in correct working order and report a dispute to the bank immediately. Here are steps that you can take:

1. Never share your password with anybody and never pass your ATM Card to any person. This would enable safety at the deposit machine.
2. Be sure that you do not count the cash in front of others.
3. Make sure that you keep your ATM Card safely, though you you can these days deposit money without an ATM Card.
4. Do not forget to take the ATM slip with you. This would serve as a proof that you have deposited the cash at the ATM.

6.2.5. Pay insurance premium

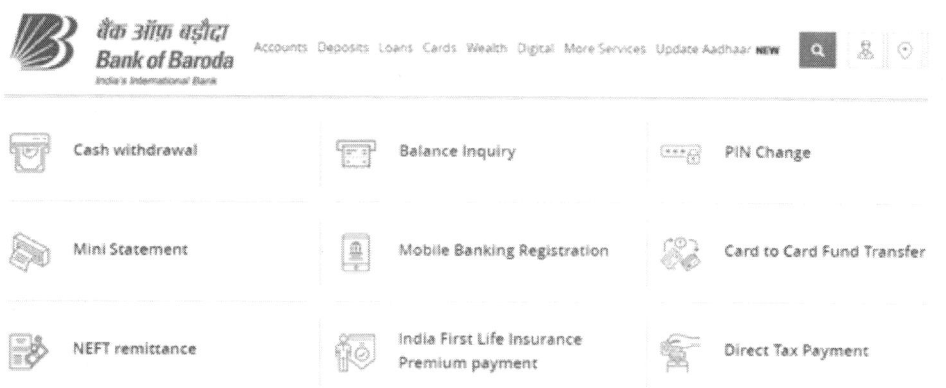

Fig Paying Insurance Premium

Insurers like LIC, HDFC Life and SBI Life have tied up with banks to facilitate premium payment through ATMs. Just keep the policy number handy. ATM facility for payment of premium is absolutely free for all customers. No any charges are to be paid to Insurance / Bank for this facility.

Procedures:

- ❖ Customer has to register his policies through the bank website.
- ❖ Registration data as per the required format/mandate from the bank is to be submitted to the bank.
- ❖ Bank sends the registration data for all the policies newly registered on next working day to PCMC in required format.
- ❖ Policies with Mode of payment i.e. Yearly, Half-Yearly and Quarterly Mode of payment are allowed through ATM payment.
- ❖ Policy should be in-force at the time of registration.
- ❖ ULIP policies and Health plans are not allowed under this scheme.
- ❖ If premium is already paid through any other mode before receipt of payment through ATM or the premium amount does not match with premium

required, the payment received through ATM goes to refunds and amount is refunded by direct credit to bank account which takes 15 days.

Steps:

- Go to 'Bill Pay' option
- Select insurer
- Enter policy number and date of birth or mobile number
- Enter Amount
- Click OK

6.2.6. Apply for personal loan

If you have an account with good bank and you need a loan, then it can be done by visiting the nearest bank ATM. Nowadays the bank has come up with a facility in which it will make available personal loans to its customers from its Automated Teller Machines (ATMs) present across the country.

The loan amount will range between Rs 1 lakh to Rs 15 lakh and the amount of interest from 11.5 to 15 per cent will vary, depending on track record of accounts. For this facility, the bank will use a combination of artificial intelligence (AI) and analytics that will analyze the savings accounts database to determine loan eligibility. The company will programme the machines in a way that it would flash an option of loan on the screen for eligible customers who use it. There will be no form filling or manual processing.

To avail of small ticket personal loans, you don't need to approach a bank branch or speak to a phone banking executive. Some private sector banks now offer pre-approved personal loans for customers using ATMs as the point of disbursal. The loan amount is decided using advanced analytics, which take into account the customer's transaction details, account balance, salary credits, and credit and debit card repayments.

For instant disbursal process for personal Loan, follow the steps at the ATM machine:

Step 1:

Customer is required to complete the primary transaction at the ATM Channel:

- Balance Enquiry
- Cash Withdrawal
- Ultra-Fast Cash Withdrawal

Step 2:

A personal Loan pre-qualified offer message is shown to the customer with the maximum offer amount a customer is eligible for.

- Click on Apply Now to proceed further

Step 3:

Next screen will show all eligible offers to the customer

- ❖ Customer is required to select any of the offers*

Step 4:

Next screen displays standard MII which is applicable to all customers

- ❖ Customers are required to accept these Terms & Conditions

Step 5:

As a two-factor authentication to initiate the disbursement of loan, customers are asked to authenticate via their Debit Card PIN

- ❖ Post authentication, the loan amount will be credited to the customer's account

6.2.7. Transfer cash

If you do not have access to Netbanking, funds can be transferred from your bank account to that of another bank using an ATM.

- ❖ You need to register the beneficiary account online or at bank branch.
- ❖ Upto Rs 40,000 can be transferred to the beneficiary account in a day, with no limit on the number of transactions.

ATM-to-ATM cash-transfer

An ATM-to-ATM cash-transfer service theoretically could attract millions more users to the machines. The service also could draw financial institutions into a self-service market that is dominated by nonbanks.

But cash transfers using multiple ATMs involve far more complex transactions than simple cash withdrawals. They involve two customers and two ATM acquirers, which means that two ATM deployers must be compensated for the transaction instead of one.

Follow the steps to transfer cash from ATM:

1. As usual, you have to first punch the SBI Debit cum ATM card at any ATM center.

Fig swipe atm card

2. Now select the preferred language which is displayed on the screen.

Fig select language

3. It asks for the PIN, and you provide the exact one and click "Yes".

Fig enter atm pin

4. Go with the "transfer" option available through the descending line-up. You can find it generally at the bottom part of the ATM display.

Fig transfer option in atm

5. Go for "Card to Card Fund Transfer" option.

Fig card to card fund transfer

6. Now provide the Debit Card number of the recipient. Generally, the number comprised of 19 digits.

Fig enter debit card number

7. You have to provide the Debit Card Number of the receiving person twice.

Fig enter amount

8. Enter the amount of money you want to transfer to the recipient as it asks and clicks "correct".

Fig confirm amount

9. Now the ATM machine asks you to confirm the amount that you are sending. Click "Yes" to confirm it.

Fig selecting account type

10. Now it asks about the account type. You have to go with the account type as its existing type; whether Savings or Current. Just select saving account and your transaction will be completed. Now collect your cash and ATM card. Don't forget to collect receipt slip. Certain cases involved with the transaction:

Fig transaction complete

That's all about the necessary steps. However, leave the place only after transaction completion message appears at the display of the ATM. The exact amount of money is channeled as you provide. You can witness it as well through the Credit from your accounts.

6.2.8. Paying bills

You can pay utility bills like your telephone bill, electricity bill, gas bill and others through ATM machine.

Paying Mobile Bill through ATM

The fastest way of paying the Bill is through ATMs. For this, all you need to know is the amount due and the bill number. The payment you made at ATM gets credited to the account within 24 hours. The payment is entirely fee-free, even if the ATM does not belong to home bank.

Fig Paying Mobile Bills

Steps:

- ❖ Find a logo of your bill payment company on the ATM screen.
- ❖ Insert your credit card into cash dispenser and select the language
- ❖ Then enter your PIN
- ❖ Select the option Mobile Services and provider.
- ❖ Click on respective service provider for payments.
- ❖ Enter the phone number you want to pay the amount due for
- ❖ Enter the amount rounded to integer crowns
- ❖ Enter the phone number you want to receive the confirmation of payment to
- ❖ ATM will print you a document with the transaction number for your record
- ❖ The confirmation about the payment done is send through SMS message on successful completion of transaction.

Credit Card Payment through ATM

There are many benefits of having a credit card. It is a convenient way of shopping, and you can credit cards are safer to carry than cash. Also if you handle your credit card wisely, you will get a good credit rating which will be useful when you wishes to apply for a loan.

Now most of the banks have the facility of making credit card bill payments through ATMs. However, you may not be able to pay your credit card bill through another bank's ATMs. You will have to use the ATM of the bank where you hold a credit card account.

- ❖ Click on Bill Pay option on the screen
- ❖ Insert your Debit Card
- ❖ Type 16 digit card number

Fig Inserting PIN

- ❖ Enter your PIN
- ❖ Enter the due amount
- ❖ The amount will be deducted from your account, and your CC bill will be paid

Payment of Utility Bills at ATM

With the CCB debit and credit cards you can pay periodic bills from a cash dispenser immediately without prior registration or paying a visit to the bank.

- Select "OTHER SERVICES"
- Select "PAYMENT OF BILLS"
- Select relevant trader
- Enter the customer number
- Select whether you want a receipt
- Screen showing customer number, amount to pay and disposable balance on account before payment.
- If you wish to make the transaction, select "PAYMENT".
- From "Payment of bills" menu, pay with debit card, Maestro and credit card by entering customer number.

6.2.9. Booking Railway Tickets through ATM Card

Fig Booking of Train ticket through ATM

We can use ATM for booking railway tickets besides withdrawing money. In this process first of all we insert our ATM card on the machine and insert our pin code in it. After inserting the ATM pin, we can perform according to our wish as indicated in the screen. As we enter ATM pin there will be options on screen:

i. Banking

ii. Railway Reservation or Railway Ticket

Now choose from the two options aforementioned. If the customer wants banking, he can click to the Banking option and proceed further while if the customer wants to book ticket, then one has to click the "ticket booking" option.

Now, as the customer clicks for the "book ticket" option, then there appears two options:

a. General Ticket

b. Reservation Ticket

Booking General ticket

If the customer presses the General ticket button then he can book a general class ticket by:

- Inserting place (both the originating place of travel and destination place)
- Inserting date of journey

In this process, the charges i.e. cost of ticket, bank service charge will be decided by bank and railway ministry. This ticket can be cancelled anytime. However the date of cancellation must not exceed the date of travel. Or the customer can cancel the ticket at the place where the journey is to begin

Booking Reservation Ticket

For booking the reservation ticket:

- enter the names of places (originating place and destination place)
- date of travel
- details of seats and availability
- inserting customer names through keyboard
- after completing all procedures, reenter the pin code to transfer the money

In this way reservation is done. This ticket can also be cancelled anytime anywhere. If needed, the cancellation can be done through ATM. The usefulness of using ATM machine can be seen during to booking of "WAITING" Ticket. The ticket issued by ATM can be used for travelling in any trains as long as the ticket is of that train and the ticket is non AC. This is due to the fact that waiting AC ticket does not permit you to enter and travel in the AC compartment

6.2.10. View account balance through ATM

After you enter your PIN, press the button on the ATM that indicates you can ask for your account balance. Most ATMs will ask you to identify whether you are asking about your checking account or savings account balance. And certain ATMs may ask whether you want the available balance displayed on the ATM screen, or just printed on the ATM receipt.

Depending on the ATM, sometimes the receipt will print a "ledger" balance and then an "available" balance. The available balance is the amount your financial institution indicates that is available to you for debit transactions such as cash withdrawals or purchases. The ledger balance, which is often a higher amount, includes items such as recent check deposits or a pending payroll credit that has been recorded by your financial institution, but has not yet cleared as good funds available to you.

Step 1 Get your ATM/debit card

Make sure you activate it and know your personal identification number (PIN) before going to the ATM. In some cases, it will activate the first time you place it in the ATM.

Step 2 Insert your ATM card into an ATM machine

You can avoid fees by using your bank's ATM, but most machines will allow you to check your balance for free.

Fig Inserting ATM card

Step 3 Type the card PIN on the machine

Your "PIN" is the personal identification number having four-digit which is given to you when you have opened your account. Type the PIN and look for the menu which give options. Scroll through these options until you find one that includes the word "balance."

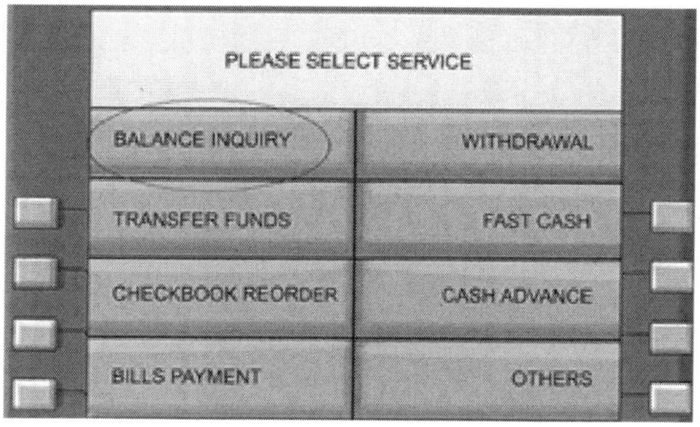

Fig Selecting the options

Step 4 Select the option to view your balance

Then, you will be able to go back and make a withdrawal or print that balance.

Step 5 Opt for receipt

Your checking balance will be printed onto your receipt for your records.

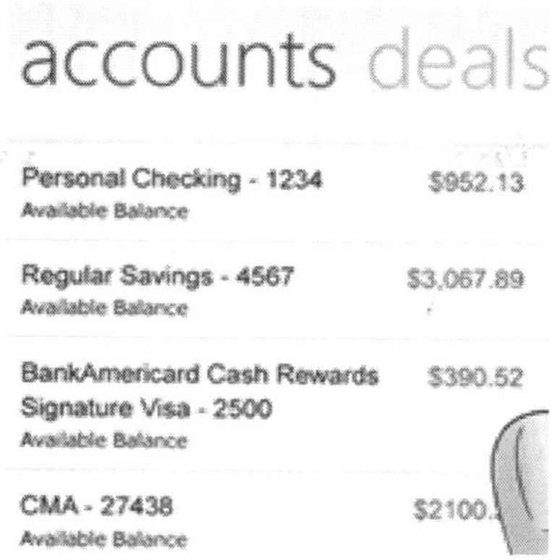

Fig Account Balance Receipt

6.2.11. Get mini statements from ATM

Mini statements are the shortened version of what your bank sends you after every three months. That is your bank account statement in short. The bank can also provide you with a printed passbook detailing all the transactions you have entered into in a particular month.

Steps:

❖ Insert your ATM card in the ATM machine

Fig Inserting ATM card

❖ Choose your language

Fig Select Language

- ❖ Enter the 4- digits secret Pin number and Press ACCEPT

Fig entering PIN

- ❖ Press "Other transaction" button
- ❖ Press "Mini Statement" button option

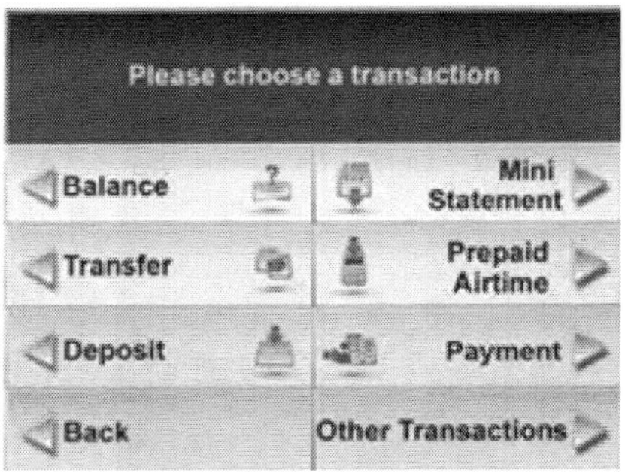

Fig Selecting Mini Statement Option

- ❖ Press "NO" to finish transaction
- ❖ Take out your ATM Card and receipt from the machine

Fig Taking Mini Statement

6.2.12. Order a cheque book through ATM

Cheques are required for withdrawing money from accounts or for making payments on local purchases. When you open a new account in any bank, you get a cheque book by filling up a form. You need to make an extra payment to get your cheque book. However, in some cases the account holders do not prefer to get a cheque book. If you have an ATM Cum Debit Card you can easily apply for cheque book through any bank ATM.

Steps:

Step 1 : Put your ATM Cum Debit Card in ATM and Enter your PIN.

Step 2 : Select the Services Option

Step 3 : Select the Request a Cheque Book Option

You will be asked if you are sure you want to request a cheque book.

Step 4 Confirm your request

In such case, the Cheque Book will be delivered within 2-3 weeks at your address what is registered with your Bank.

6.2.13. Making donations through ATM

People are now able to donate money to charities including Cancer Research institutes at the same time as withdrawing cash at thousands of ATMs. More than 8,000 cash machines across the globe, have been enabled to accept donations between £1 and £150 for charities.

Donations will be offered as a separate menu item on the ATM screen, or as a post-transaction option, so that making a donation will not delay those who just want to withdraw cash or check their balance.

- Tesco Bank said people could donate to its chosen charities at ATMs at more than UK 2,200 stores. People do not need to be RBS or Tesco Bank customers to use their machines to donate money.
- RBS cash machines will also allow donations to the STV Appeal, which helps children and young people in Scotland who are affected by poverty.
- HSBC already allows its customers to donate to charity through its cash machines, in an initiative that has been operating since 2005. HSBC said £341,623 was donated in 2010, despite the general squeeze on consumers' budgets.
- SBI allow customers to donate for charity in (a) Vaishno Devi (b) Shirdi Saibaba (c) Gurudwara Takht Saheb, Nanded (d) Tirupati (e) Sri Jagannath Temple, Puri (f) Palani, Tamil Nadu (g) Kanchi Kamakoti Petam, Tamil Nadu (h) Ramakrishna Mission, Kolkata (i) Mantralaya, Andhra Pradesh (j) Kashi Vishwanath, Banaras (k) Tulja Bhavani, Mumbai (l) Mahalakshmi, Mumbai and many more.

6.2.14. Getting the currencies through ATM

An ATM will normally give you a better exchange rate than you'd get if you were buying foreign currency with cash or traveler's checks, since the exchange is made at the interbank wholesale rate. But watch out--some banks are now charging "conversion fees" on top of the interbank commission.

Although some banks have high fees to use foreign ATMs, not to mention adding on high foreign-transaction fees, the ATM is still almost always the cheapest option for changing your money. And if your bank has international ATMs or partner banks abroad, you can sometimes save a little on your cash withdrawals; that's true even at BOA, which charges just 1% at member banks and no other fees. Capital One Bank charges nothing for a foreign ATM withdrawal, while other banks such as Chase charge 3% on top of every withdrawal as well as a $3 fee. It pays to shop around for a bank if you travel a lot internationally.

The currencies at each ATM can differ depending on the location, but the common ones are:

- Japanese Yen
- US Dollar
- Great British Pound
- Indonesian Rupiah
- Euro
- Singapore Dollar
- Malaysian Ringgit
- Thai Baht
- New Zealand Dollar
- Papua New Guinea Kina
- Australian Dollar

Getting Currency in Abroad

Whenever you use your card abroad, most POS machines and ATMs can automatically pick up the country where your card has been issued. Once it has

figured out that your card is 'foreign', the machine assumes that you don't hold the currency it can spit out and will offer you a seemingly innocent question that reads something like this:

The small message above confirms that the ATM will only pay out in the local currency, regardless of what currency you choose to debit your account in. It also highlights that a surcharge of 6% is included in the rate displayed on screen and gives an exact GBP cost for taking out 200 Euros. Despite showing a fixed amount in a currency you'll be natively familiar with, you should be wary of the machine's offer, since it is openly admitting to adding a 6% fee to the exchange rate offered.

If you choose this option, the merchant or ATM provider can apply their own exchange rate. This rate usually has a mark-up on the interbank rate (in this case 6%) enabling the ATM provider or merchant to make a profit on your transaction.

Debiting your account in the local currency (EUR) might seem like the less desirable option since you are not shown the equivalent rate in GBP, but this is all part of the sneaky plot to take some extra money from your pocket.

By choosing this option, the ATM will charge your card in Euros, but more importantly, it will let the card issuer handle any currency exchange, meaning you will benefit from using your bank's exchange rate or, in Revolut's case, the interbank exchange rate - no mark-ups or hidden fees.

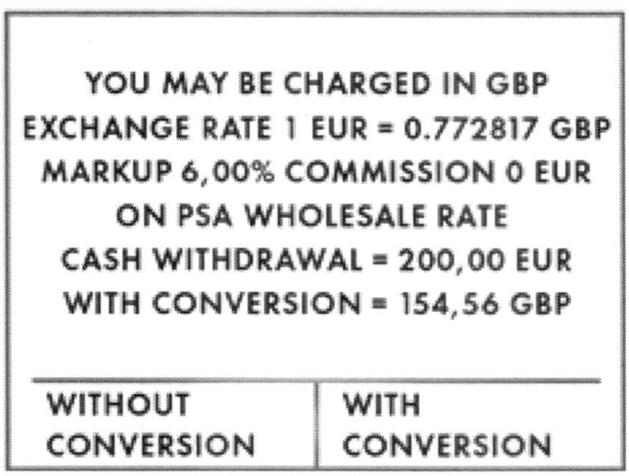

6.2.15. Change of ATM Pin through ATM

Bank customers can change 4 digit PIN number of their debit card at ATM counters. This facility is already provided by all ATM centers. This can be done at regular intervals. Changing a debit card's pin number issued by the bank can be change easily through ATM machine. One doesn't have to visit a bank branch because the pin changing operation could even be done at an ATM machine as well. Changing the ATM once in a while is in fact a good practice, but remembering the new PIN every time is even more important.

A pin number is issued with each ATM debit & credit cards for more secure transaction. A 4 digit pin number would mean a lot of possible combinations but the ATM machine won't allow more than 3 in most cases, so a simple PIN code is a lot of security for a bank account.

To change the pin number of your ATM card, you need to use same bank ATM machine whose ATM card is with you. Although money withdrawal is possible from other vendor's machines, pin changing operation is only enabled in the respective bank ATM machines. For this you have to visit your bank ATM machine and do the following :-

1. Insert or swipe the card in your ATM machine. Wait until the machine accepts your card and ask for the PIN number.

Fig Insert the atm card

2. Now Choose desire language from the list.

Fig select language option

3. Enter your existing pin and when the menu appears on screen.

Fig enter existing atm pin

4. Next click on the "banking" option displayed on screen.

Fig Selecting banking options

5. Inside the Banking option, there should be few options. You need to find 'Pin Change' which should be located somewhere easy to find.

Fig Selecting pin change option

6. Tap that option. You might be asked to enter your current PIN again and then the new PIN number you want to set. You need to re-enter new four digit pin number.

Fig Entering new pin number

7. After you confirm your action, the PIN number will be changed. It will show message "your pin has been changed successfully".

Fig Successful PIN change option

Once you are done with this operation, you will have to use the new PIN number every time you use the card in a machine.

6.2.16. Mobile banking Registration through ATM

Mobile banking through cell phone offers many advantages which helps account holders to manage their bank account(s) on their smart phone device, at their convenience. Here are the advantages of Mobile Banking;

- Anytime banking: Mobile banking gives you the privilege of banking anytime and anywhere.
- Mobile Banking is free: The service provided by bank is free of charge. Various banking services provided include Account Balance Inquiry, Debit Alerts, Bill Payment Alerts, Transaction History, Fund Transfer Facilities, Minimum Balance Alerts etc. can be accessed from your mobile. You can transfer money instantly to another account in the same bank using mobile banking.
- Secure Banking: The app comes with advanced technologies making it safe and secure as Internet Banking.

Mobile banking helps the customer to do following transactions:

- Balance Enquiry
- Mini Statement
- Funds Transfer within Bank from one account to another
- Funds Transfer to accounts of other bank using NEFT and IMPS
- Topup and bill payments
- Stop cheque
- Cheque status
- ATM card blocking
- Foreign Inward Remittance using IMPS

Registration Steps:

- Walk into the Bank's ATM to register Mobile Banking Service
- Insert/dip the ATM card in the ATM machine
- Now on prompting, enter ATM Card's PIN
- Selects Mobile Banking Registration option in ATM Menu
- Enter 10 digit Mobile number and select Confirm button
- Re-confirm the Mobile number and Confirm
- Select the Account Type options <CURRENT> or <SAVINGS>
- On validation, ATM displays the message "Your Mobile Number Registered successfully" and will print the message
- If customer details found invalid, then ATM displays "Registration Failure".
- After successful registration, Customer receives SMS on Mobile from Bank regarding Mobile Application URL link for downloading Mobile Application Software
- Select two passwords 1) Application password and 2) Activation code/mPin

6.2.17. Paying Credit card bill through ATM

There are many benefits of having a credit card. It is a convenient way of shopping, and you can credit cards are safer to carry than cash. Also if you handle your credit card wisely, you will get a good credit rating which will be useful when you wish to apply for a loan.

Now most of the banks have the facility of making credit card bill payments through ATMs. However, you may not be able to pay your credit card bill through another bank's ATMs. You will have to use the ATM of the bank where you hold a credit card account.

To make the payment:

- Insert the ATM card
- Enter the credit card number

Fig entering credit card number

- Choose the amount you want to pay
- On entering the required amount, the account that is linked with ATM card will get debited

Chapter 7
Maintenance and Cleaning of Automated Teller Machine (ATM)

7.1 Introduction

Maintenance is a way to catch small issues with your machine before they become big issues. It involves thorough cleaning to remove dust from inside your machine. If the dust is not removed, it can cause your machine to overheat and have premature failures. Maintenance is needed to help guarantee the accuracy of your equipment and to extend the life of your machine. When your equipment is under a preventative maintenance program, the equipment not only runs better but also enhances the capabilities of the machine. It gives the service technician the opportunity to ward off any potential problems that may occur under normal wear and tear of your machine. Under a preventative maintenance program, your equipment is cleaned, inspected, adjusted or updated, and tested by a professional. A highly knowledgeable technician that truly cares about the accurate operation of your equipment will make sure these are completed.

Fig ATM maintenance

7.2 Types of ATM Maintenance

7.2.1 Corrective Maintenance

The careful reader will note that the requirement that "Entries will be made in the log when the ATM is started up and shut down" has not been reflected in any of the diagrams or the code. What changes would be needed to implement this requirement?

When the system asks the customer a question, it waits as long as necessary for the customer to answer. This could cause a severe security problem if the customer were to leave at this point without answering the question, since the ATM card would still be in the machine, generally with the customer's PIN having already been entered, thus allowing another user to "hijack" the session and perform operations like cash withdrawal from the customer's account.

Fig ATM Corrective maintenance

While this seems unlikely, it is conceivable in some cases that a distracted customer might make this mistake. For example, after performing a transaction, the system asks the user "Would you like to do another transaction?". If the customer has just withdrawn cash, it is conceivable that the customer might place the cash in his/her wallet and accidentally forget about answering the question and retrieving the card.

A real ATM typically has a timeout mechanism that begins beeping and ultimately ejects the card and ends the session if the user fails to respond within a certain amount of time. While a customer who leaves before this happens might inadvertently leave a card behind, at least the customer would not be "logged in" with his/her PIN. (Thus, the security risk is comparable to that resulting from a lost card, with the customer's account still PIN protected.) What changes would be needed to the system requirements, design, and code to improve security this way?

7.2.2 Adaptive Maintenance

Many ATM's have cash dispensers that can handle two different denominations of bills - often $20 and $5 - which allows a user to request cash in any amount that is a multiple of $5. Suppose the cash dispenser in the example (which only handles $20 bills) were replaced with such a dispenser. What changes would be needed to take advantage of the enhanced capabilities of such a dispenser?

Fig Adaptive maintenance

7.2.3 Perfective Maintenance

When the customer's card is retained, an entry should be made in the log to this effect. Many ATM's offer the customer additional options beyond the basic four listed here - e.g. making a payment on a loan the customer has from one of the accounts linked to his/her card. What changes would be needed to make it possible for a customer to make a loan payment at the ATM?

7.3 Proactive ATM Maintenance

When an ATM fails, it can cost the operator a significant amount of time and money, not to mention the damage done to the operator's reputation. With social networking apps at their fingertips, customers who find an ATM they depend on out of service can tarnish a brand with just a few keystrokes.

To avoid these headaches, financial institutions and third-party providers of ATMs should consider a move from reactive to proactive maintenance. Reactive maintenance — too often the default — deals with problems only after they occur.

Proactive ATM maintenance, on the other hand, focuses on identifying and solving problems before they materialize.

But in order to ensure that proactive ATM maintenance is worth the investment, ATM providers must test the results to ensure they can accurately and dependably prevent failures.

As previously mentioned, reactive maintenance of ATMs can require a considerable investment of time and money.

If the organization isn't aware of a problem until it occurs, it might be forced to dispatch an engineer to the scene, which is costly and disruptive to the regular schedule.

And, of course, the longer an ATM is down, the more business the provider loses. Consumers have grown increasingly impatient with slow and inconsistent self-service experiences, and become easily frustrated when the ATM they visit is down.

Given these facts, taking a proactive approach to ATM maintenance has some advantages over the reactive model that are worth considering.

One type of preventative maintenance is typically performed by the engineer who has been sent to an ATM in response to problem. While the engineer is at the ATM, he or she will carry out preventative maintenance in an attempt to prevent future issues.

Engineers generally will check the top 10 known fault conditions while they're at an ATM, but their schedule is usually too full to conduct many additional tests that might be useful.

Fault testing can be a largely manual process requiring a good deal of time, however it's certainly better than no preventative maintenance at all.

Another approach to proactive maintenance might be more attractive to organizations that wish to improve efficiency while lowering operational costs.

This approach uses a monitoring agent onboard the ATM that automatically screens the behavior of the device and records errors and warnings. When a fault is detected, the monitoring agent can execute local remedial rules to attempt a fix. If the problem cannot be resolved in this way, the system can forward the events to a central monitoring server for action.

The monitoring server analyzes data received to identify any patterns that might have led to the failure. By working backwards through the data to determine the sequence of events that occurred at the ATM before the failure, the system can potentially recognize the symptoms of a particular type of failure.

If the system is aware that a pattern of X, Y and Z warnings might lead to error A, it can apply a rule set for a potential situation and alert the operator to the need for preventative action — for example, to replace a part that is expected to fail soon.

Using this approach means conducting maintenance on machines when they need it. The automated and adaptive nature of this approach also means that organizations can easily and quickly respond to changes in a machine.

The problem most ATM providers encounter with this approach is that the only way to verify that an event or pattern indicates a particular outcome is to manually test the machine in the lab using data gathered from machines in the field.

This is not only burdensome, but also time consuming, as it might take two to three months to thoroughly analyze the data. Only then can the operator accurately determine which errors can be avoided, how often a rule works, and how often the error still occurs despite intervention.

From there, the tester can make a recommendation to test the rule again, or to test other conditions instead.

To eliminate the weaknesses in a traditional, lab-based approach, ATM monitoring rules should be tested in a virtualized hardware environment, where countless complex failure scenarios can be creating using scripts.

A script can run different events, recreate complicated situations and test how the monitoring solution detects these events, executes the rules and maintains the ATM.

Proactive ATM maintenance and monitoring can offer a clear competitive advantage to an organization that manages ATM networks and wishes to avoid the cost, expense and reputational damage associated with unplanned outages.

Using hardware virtualization and test automation, an organization can develop and implement sophisticated monitoring rules to ensure that ATM availability remains high and ATM users remain satisfied.

7.4 Cleaning of ATM machine

Cash machines/ ATMs are used daily by millions and millions, making it the perfect surface for germs to spread. It's not very often you see a shiny ATM keypad. It is a real shame as they are often the first point of contact for many customers. The experience must be safe, healthy and representative of the brand. The absence of cleaning also leads to the malfunction and failure of machines.

Cash machine cleaning includes

- ATM Keyboard (Touch pad/keypad)
- Screen
- Surrounding
- Enveple holder
- Signage
- Card reader
- Lights
- Camera
- Receipt dispenser
- Safe anti bacterial solution (kills 99.9% of germs)

Fig uncleaned pad

Branch transformation initiatives have put the ATM self-service channel front and center as a means of routinely interfacing with customers. ATMs have become a multi-functional banking device where consumers can perform virtually any kind of transaction. This makes ATM uptime a critical component of the successful customer experience. Over 30% of ATMs worldwide now include crucial automated deposit modules - and that is only increasing. These highly efficient units are continually exposed to the dirt, ink and contaminants typical of currency and checks, potentially slowing down transactions, triggering errors or causing the units to jam. Card readers, touch screens and key pads are also subject to extended use, finger oils and environmental conditions. Implementing a proactive cleaning regimen is proven to maintain the operational uptime of your ATM fleet.

7.4.1 Steps in cleaning of ATM

An unclean ATM may deter customers from coming back, as well as increase the chances of it malfunctioning. Follow these three helpful tips each spring to get your ATM clean, working and looking as good as new. Before you begin, always be sure to turn the power off and unplug all electrical cords, and never use water on any electrical components.

1. Begin your ATM and ATM equipment spring cleaning by removing debris from the bottom of the vault. Use a vacuum to suction out any dust or small particles that could kick up and cause jams in the motor or sensors. Wipe down the vault with a damp cloth to remove any built up grime on the walls.

2. Use a can of compressed air to clean the cash dispenser and printer assembly. This will remove any accumulation of dust or paper pieces that could be affecting sensitive parts, such as sensors or roller components.

3. Create a mild cleaning solution by combining two drops of dish soap per 12 ounces of warm water. Use this solution to wipe down the pick wheels, belts, and outside surface areas of the ATM. Be sure not saturate any components, as too much water can cause corrosion and other damage.

4. Use dry lint-free cloth to wipe up the surface of the interior.

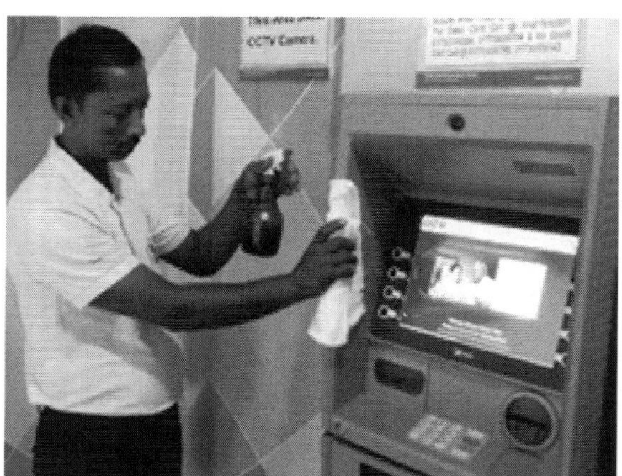

Fig Cleaning of ATM with dry cloth

5. Use a lint free cloth dampened with cleaner to clean the paint or grease on the spray powder areas of the interior. Using a lint-free cloth dampened with alcohol to clean the paint or grease on the plating parts of the interior.

These simple cleaning tips will ensure the ATM is functioning at maximum efficiency and providing your customers the best possible experience. Depending on your ATMs location, it may be necessary to perform these measures several times throughout the year. Its important to have a clean ATM to keep customers returning.

7.4.2 Tips for keeping ATM clean

1. Keeping Dispenser Warm

When the temperature drops below 45 degrees Fahrenheit, your ATM's tension belt and motor may begin to have operational issues. By keeping a small electric heater that's approved for tight spaces inside the vault, you can keep your machine warm and running smoothly. It may be a good idea to put the heater on a timer so it cycles on and off every few hours. This will maintain the temperature inside the vault and prevent other components from overheating.

2. Keeping ATM Proper Covered

Outward facing ATM components such as the keypad and function keys have small gaps that water can sometimes enter. Even if the machine is intended for outdoor use, condensation and water droplets can occasionally get in and damage the equipment. For this reason, it's always a good idea to use a kiosk designed to protect the ATM from direct snow or rain.

3. Proper ATM batteries

Most batteries don't respond well to steep temperature drops, and can sometimes fail to work when you need them the most. If you use an ATM electronic lock to open the vault, make sure you always have access by frequently changing the batteries - even if they still work. You can do this by keeping several battery sets in your car and cycling them out on a regular basis throughout the winter months. Doing this will eliminate battery failure before it happens, and ensure you always have access to your vault.

4. Proper Sheltering

Always keeping your ATM properly sheltered is a good tip, but winter storms can bring unpredictable amounts of snow, sleet and wind. If a winter storm is on its way, prepare ahead of time to prevent major damage to your outdoor ATMs. It may even be worth placing the ATM out of service and covering it with plastic until the harsh weather has passed. This may be a small inconvenience now, but could reduce costly downtime in the future due to a component failure.

7.5 Cleaning of ATM hardware parts

7.5.1 Cleaning head of card reader

- To clean the head of the card reader you will require a cleaning card. First, apply a little cleaner on the cleaning card, in normal mode, when the monitor displays:
- Please insert your card, proceed as follows:
- Insert the cleaning card.
- Wait for the card to come out.
- Repeat the procedures 10 times.

7.5.2 Cleaning of air filter

Fig Air Filter

- Cleaning method: Take the air filter out of its holder by pulling it upwards, wash it with water and dry completely, then restore the air filter.
- Cleaning period: Cleaning period changes with the ATM operation environment, service engineers should inspect the air filter periodically, and wash it every one month in areas with heavy dust or every half year in areas with light dust.

7.6 Cleaning Equipment required for ATM

Fig ATM cleaning materials

The cleaning kit contains recommended materials for servicing and cleaning your ATM, the materials are shown as follows:

- General surface cleaner
- Lint-free cloth
- Card reader cleaning card

Cleaning articles used are:

- lint-free cloth
- general surface cleaner
- dilute detergent solution.

Chapter 8
Customer usage of Automated Teller Machine (ATM)

8.1 Introduction

Automated Teller Machine (ATM) refers to a machine that acts as a bank teller by receiving and issuing money to and from the ATM account holders/ users. The evolution of ATM was not in isolation, rather as a result of the general global wave in the technological revolution. This came due to the need to respond to the challenge of the multiple bulk of daily complex information that arises from among others; increase in competition, increased customer demand for both service provision as well as efficiency, expansion due to the increase in demand for services etc.

The introduction and use of ATM system of banking therefore brought in efficiency in the banking industry majorly in terms of speed, data processing and storage. Thus it saw a huge improvement in the long queuing in most banking halls. Despite of all the merits of the ATMs, customers still complaint of shortcomings on the use of the system like; break downs of ATMs, long queues at ATM service points, retention of customer's cards. Thus the researchers intends to find out why still these cries.

8.2 Background

The history of banking is that till the industrial error, banking services were rendered on manual basis characterized by; ledger keepers at back office, cashiers providing front desk service, ledger cards, cash registers among others. This is not only tedious, but also strenuous and slow besides providing inaccurate and unreliable information on a regular basis because of the human error. The limitations of manual banking highlighted above in most cases results to customer complaints, slowness in service provision, long queue and high operational costs etc. These limitations reduce reliability of financial services as regards to accuracy and completeness.

ATM is the abbreviation of Automated Teller Machine which acts as a teller in a bank who takes and gives money over the counter. ATMs are placed not only near or inside the premises of banks, but also in locations such as shopping centers/ malls, airports, grocery stores, petrol/gas stations, restaurants, or any place where large numbers of people may gather. ATM services includes function such as cash withdrawal, balance enquiry, bill payment, cash and cheque deposit, saving and credit account on a 24 hr basis. Thus with the appearance of ATMs, some limitation of time and geographic location has been resolved.

8.2.1 ATM system

The introduction and use of ATM system of banking has received different perceptions. One of the views is that, it may not have really created customer satisfaction for bank clients, and the other is that, it may have. Despite of all the merits of the ATMs, customers still complaint of shortfalls on the use of the system such as; break downs of ATMs, long queues at ATM service points, retention of customers cards, limited knowledge on the use of ATM cards, fraudulent transactions and its operation in just a few languages.

It's therefore against this background that the researcher was prompted to investigate and find out the effect of ATM system of banking in enhancing customer satisfaction in CERUDEB, Gulu branch.

ATMs typically connect directly to their ATM Controller via either a dial-up modem over a telephone line or directly via a leased line. Leased lines are preferable because they require less time to establish a connection. It is observed that, most modern ATMs, the customer is identified by inserting a plastic ATM card with a magnetic stripe or a plastic smartcard with a chip that contains a unique card number. Security is provided by the customer entering a personal identification number (PIN). For one to access ATM service, he/ she (the card holder) has to insert the card (magnetic strip card) into the machine (ATM), which then reads the strip and makes contact with the central computer to confirm the genuity of the card which is either accepted rejected depending on whether it is valid or not. When accepted, the customer then punches his/ her PIN number which is then verified according to its compatibility with the information stored in the card. After which it then perform the service requested of like (issuing cash, accepting cash/ cheque deposit, balance enquiry, mini-statement) etc, and finally ejects the card.

Automated teller machine (ATM) is an innovative service delivery mode that offers diversified financial services like cash withdrawal, funds transfer, cash deposits, payment of utility and credit card bills, cheque book requests and other financial enquiries (Muhammad 2010). The concept of automated teller machine (ATM) has been quite old and has been developing throughout the world. Undoubtedly, a fair number of theoretical and empirical researches have been made worldwide, because of increase in scale of ATMs and consequently to convert the economy into a cashless one.

ATM is commonly use by customers of various banks for cash withdrawal and account balance enquiry. The other functions of the ATMs such as funds transfer, cheque-book requests and payment of utility bills among others which are collectively aimed at increasing customers' satisfaction are not fully utilized. This inability on the part of customers to exploit these opportunities provided by the ATM is mainly due to reasons such as inadequate knowledge about the functions of the ATMs apart from cash withdrawals and balance enquiry, inability of some banks to render diversified ATM services, among others. However it should be noted that ATM services offered to clients differ among banks.

Mcandrews outlined that; the various utility of ATM's has gained worldwide popularity. The utilities include withdrawal of cash as per convenience of the customer than during the banking hours of various branches. Besides providing off-time and off-shore services, there is reduction of cost servicing. Ones access is gained, it offers several retailed banking service to customers. They are normally located outside the banks and also found at airports, malls and places far away from the home bank of customers. They were introduced first to function as cash dispensing machines; however, due to advancements in technology, ATM's are able to provide a wide range of services such as making deposits, funds transfer between two accounts and bill payments. Banks tend to utilize the ATM as all others, for competitive advantage. ATM's are a cost–efficient way of yielding a higher productivity as they achieve higher productivity per period of time than human teller (an average of about 6400 transactions per month for ATMs compared to 4300 for human tellers(ibid).

Stuart (1999) pointed out that the total number of ATM transaction have more than doubled over the last ten years and its estimated to reach near eleven million and total number of terminals in the united states has tripled over the last ten years.

Several empirical studies have being carried out on ATM services and customer satisfaction. Saxeam and Pahwa for instance, analyzed customer satisfaction at ICICI bank with special reference to ATMs in Udaipur city. Their study looked at some aspects such as the service quality of ATM personnel, location, sufficient number of ATM in the city, the regularity in working of ATMs, their impacts on overall performance and their options on various other related issues. Their study revealed that the quality of ATM services provided was marginally above average and hence customers' satisfaction with the ATM service was also above average.

Chelliah et al, also studied service quality delivery and its impact on customer satisfaction in the banking sector in Malaysia. His study reveals that assurance, reliability, tangibles, empathy and responsiveness contributed to 62.1% effect on customers' satisfaction. Only tangibles have significant effect on customer satisfaction and hence indicated it with a P-value of less than 0.01.

As per study; determinants of ATM service quality and its effects on customer satisfaction concluded that convenience, efficient operation, security and privacy, reliability and responsiveness have a significant dimension on ATM service quality. Also, in a study on customer satisfaction of ATM services, study investigated the satisfaction levels of ATM card holders with respect to various aspects of using ATM and their opinions on various other related issues. The main objective of their study was to examine the level of satisfaction of staff and non-staff customers of ATM. In their study, they categorized the various aspects of ATM as the promptness of card delivery, performance of ATM and the service quality of ATM personnel. After their survey and analysis of data, the study revealed that the average satisfaction levels of ATM users is more than that of non-staff ATM users in all respects or aspects of ATM.

8.2.2 Effectiveness of ATM

Without usage of technology the banking sector cannot provide customers with effective services. Effective service delivery is a new or significantly improved service concept that is taken into practice. Customer expectations concerning service encounter experiences and service delivery mechanisms as well as the entire concept of what constitutes quality service are therefore key issues that need to be considered prior to the implementation of any structural change. Effective service delivery is a service product or service process that is based on some technology or systematic method. It can be a new customer interaction channel, a distribution system or a technological concept or a combination of them. As observed, effectiveness of service provision have a significant relationship with overall customer satisfaction. Effective service delivery is positively related to customer satisfaction in that, when a customer perceives that the delivery mode of the transactions that the bank is supposed to offer is quite good, the more the customers will be satisfied with the bank services.

8.3 Customer Satisfaction with Technology

The evolution of information and communications technology (ICT) has brought development and adoption of advanced technologies in commerce and industry. One such modern technology in the banking sector is the Automated Teller Machine (ATM) system. An ATM system is an inter-organisational system that links banks and other financial institutions to retail banking customers for several types of routine banking transactions. These include inquiries, deposits, cash withdrawals, cash transfers and payments. Although ATMs were introduced in banking in America in the early 1970s which begins into common operations in early 2000. A decade later, all commercial banks own a network of ATMs for the delivery of banking products/ services.

Literature shows that ATM banking has received customer preference to become the second most popular channel for accessing banking products/services behind branch banking. ATMs provide bank customers with 24 hour access to banking products/services; they are easy to use and are faster than human tellers in the banking halls. ATM systems are believed to have improved the operational efficiency of banks and customer service in the banking sector.

Although ATM systems have high fixed costs, they have lower variable transaction processing costs according to research. With that proficiency ATMs could be substituted for employees that provide services on demand deposits accounts thereby be able to reduce the number of transactions processed by human tellers.

That would allow banks to reduce direct customer service employment. ATMs have therefore become strategic technology in the banking sector in delivering banking products/services and are expected to offer competitive advantage to banks investing in ATM technologies over those that do not.

Today, investment in ATM technologies remains strategic as banks continue to invest in newer and more efficient ATM technologies to bolster delivery of an efficient banking experience. It is the goal of banks to offer competitive services and keep an expanding base of satisfied customers to remain competitive and profitable. This is evident through banks' investment drive in improving and increasing delivery channels, product/service reach and customer communication. However, although ATM systems are believed to have improved operational efficiency and customer service, the study by Banker did find that ATMs did not affect the banks market share. The study on the effects of early adoption of information technology by Dos Santos and Peffers (1993) found that not all early adopters of ATM technology were able to improve human resource proficiency and the study further found that in some cases early adoption of ATM banking even decreased efficiency.

The studies on adoption of ATM systems however reveal that early adoption of ATMs only increased market share in certain environments and that the competitive advantage created by early adoption of ATMs was short lived, enjoyed only until other banks adopted ATM technology. As studies on early adoption of ATM technologies have shown that competitive advantage achieved through adoption of ATMs was short lived only until other banks adopted ATM technologies, this study was set to assess the impact of ATM banking performance on customer satisfaction with banks. Now that each bank in Malawi has installed ATM technologies to deliver banking products/services, is the performance of these ATM technologies critical to customer satisfaction with their banks? Are ATMs still an area that can offer a bank competitive advantage or are they just an investment requirement in retail banking and that banks should look elsewhere to achieve competitive advantage and customer satisfaction?

Customer satisfaction is the state of mind where customers have about a company when their expectations have been met or exceeded over the lifetime of the product or service. Increased customer expectations have created a competitive climate whereby the quality of the relationship between the customer and bank has taken on a greater significance in some cases than the product itself. Banking industry strives to succeed by putting the topic of rapid and changing customers needs to their agenda. This can be achieved through good customer care and offering attractive services or products that other competitors may not offer. Therefore, customer satisfaction is seen as a key performance indicator within business. The concept of customer satisfaction occupies a central position in marketing and practice. Customer satisfaction is a person's feelings of pleasure or disappointment resulting from comparing a product's perceived performance or outcome in relation to his or her expectations. Customer Satisfaction is the actual meeting of customer's expectation after he or she has completed consuming a product or service.

A customer is the most important visitor on our premises. He is not dependent on us. He is not an interruption to our work. He is the purpose for it. He is not an outsider on our business rather he is part of it, we are not doing him favor by serving him rather he is doing us favor by giving us an opportunity to do so. As described, a customer as one that enables the organization exists. They are the purpose of our business. Customers are the heart, the life and the soul of our businesses, without whom we can't hold even for a second to exist in business. Hence they should be accorded utmost respect and care when rendering service to them.

Khirallah defines customer satisfaction as; a customer's perception that his or her needs, wishes, expectations, or desires with regard to products and service have been fulfilled. Consumer satisfaction in short can therefore be defined as an evaluative process that contrasts pre-purchase expectations with the actual perceptions of performance during and after consumption experience. In summary, Customer Satisfaction is the meeting or even exceeding of a customer's expectation after the use product. The outcomes of satisfying a customer are:-

8.3.1 Customer loyalty-

LOYAL customers are those who have the enthusiasm about the brands or products they use. The more enthusiastic a customer is, the higher the profit contributed to the brand.

Loyalty is a combination of intentional repurchase behavior and psychological attachments of a customer to a particular service provider. The fundamental assumption of all the loyalty models is that keeping existing customers is less expensive than acquiring new ones. In summary, Loyalty is customer's demonstration of faithful adherence to an institution despite the occasional errors. Thus satisfying a customer is very paramount to organizations existence.

8.3.2 Customer retention-

Customer Retention is the ability to hold on to customers over time. Customer retention is the activity that the selling organization undertakes to reduce customer account defections. It can also be described as a series of actions that the selling organization undertakes to reduce defections. Ganesh et al., observed that, long-term, customers become less costly to serve due to the bank's greater knowledge of the existing customer and to decrease serving costs. They also tend to be less sensitive to comparative marketing activities. Loosing customers not only leads to opportunity costs because the reduced sales, but also to an increased need for attracting new customers which is five to six times more expensive than customer retention.

8.4 Measuring Customer Satisfaction

Customer satisfaction is one of the frequently researched concepts in marketing. At the centre of understanding customer satisfaction is the concept of service quality. Although there are no commonly accepted definitions of both concepts of customer satisfaction and service quality; service quality is defined as a measure of how well a service delivered matches the customers' expectations. It is said that service quality is best defined by the consumer of the service thereby making it subjective in the sense that two persons could perceive service quality delivered differently.

Literature posits that service quality is an attitude of overall judgment about service superiority. The judgment that is obtained from comparing performance perceptions with expectations or performance alone. It is based on this judgment that customers perceive satisfaction with the service. Customers perceptions are beliefs about the experienced service and expectations are beliefs about a service having the desired attributes and that expectations are used as a standard to service evaluation. Service quality therefore is an antecedent of customer satisfaction highlighting the importance of service quality in influencing customer satisfaction and customer behaviour towards brand loyalty, repeat purchases and willingness to recommend.

Measurement of customer satisfaction has been dominated by the performance disconfirmation paradigms. However, it is given in a review of nineteen service quality models that can be used to measure customer satisfaction with a product/ service and these include:

- Technical and functional quality model
- GAP model
- Attribute service quality model
- Synthesis model of service quality
- Performance only model
- Ideal value model of service quality
- Evaluated performance and normed quality model
- Alignment model
- Attribute and overall effect model
- Model of perceived service quality and satisfaction
- PCP attribute model
- Retail service quality and perceived value model
- Service quality customer value

The popular approaches to measuring customer satisfaction however are; importance – performance model which proposes that a customer's satisfaction with a product or service is a composite of

1. the perceived importance attached to the range of the product/service attributes or benefits

2. the customer's beliefs about the degree to which the product/service has each attribute. When a product/service is perceived to have and performs in the attributes customers perceive important, it is expected that customers will be satisfied. Secondly, the expectations-performance paradigm on the other hand states that a customer is satisfied with a product/service if the product/service performance meets or exceeds the customer's expected performance of the product/service. The model is based on GAP analysis in the conceptualization and delivery of the service quality. The researchers visualize the existence of service quality gaps. The first gap is the difference between customers' expectations and management's perceptions of customer's expectations, that is, not knowing precisely what the customer expects. The second gap is the difference between management's perceptions of customer's expectations and service quality specifications drawn to guide the service delivery, that is, improper service quality standards.

3. the difference between service quality specifications and actual service delivered to the customer that is the service quality gap.

4. the difference between the service delivery and the external communication to the customer about the service delivered, that is, matching service delivery and promises.

5. the difference between customer's expectations of the service and the perceptions of the service delivered. This last gap forms the perceptions of service quality in the customer, therefore service quality becomes a function of perceptions and expectations.

Based on the GAP analysis developed a customer satisfaction measurement framework popularly known as SERVQUAL. SERVQUAL is a multi-dimension and multi-attribute measurement scale where customers rate a product/service on a set of general attributes on a Likert type scale. Parasuraman identified ten service quality dimensions which were later collapsed into five service quality dimensions of tangibles, reliability, responsiveness, assurance and empathy with 22 general service quality attributes. SERVQUAL has been widely used in service quality and customer satisfaction studies across industries.

SERVQUAL has drawn criticism from Carman who argues that subjects should rate a product/service on criteria that are relevant to the specific product/service supporting empirical researchers who argue that the general attributes have potential to carry in-built bias towards dissatisfaction since expectations are determined by the ideal product/service. Nonetheless SERVQUAL is one of the popular measurement models of customer satisfaction which has been used in various studies and remains one of the popular service quality measurement models.

Cronin and Taylor eliminated the element of expectations SERVQUAL model perceived that service quality and customer satisfaction is a form of attitude that performance only is an enhanced means of measurement. This led to the development of SERVPERF model which uses the same multi-dimension and multi-attribute scale as SERVQUAL but service quality is evaluated by measuring performance only without expectations or importance weights being used as a standard of comparison. Several researchers agree that assessment of service quality and customer satisfaction may depend solely on performance and that SERVPERF may have less bias than SERVQUAL in measuring service quality.

Respondents to the study were from five commercial banks in Malawi namely; Financial Discount House (FDH) Bank, Savings Bank (SB), New Building Society (NBS) Bank, Standard Bank and National Bank (NB). All respondents have been with their respective banks for a period of over two years and 87.2 percent of the respondents use ATM banking at least every week.

The study received 71 percent of the responses from the targeted study sample where 40.6 percent of the responses were from female respondents and 59.4 percent from male respondents. The majority of the respondents (81.3 percent) were aged between 26 and 45 years and 71.9 percent had graduate to postgraduate qualifications while 28.1 percent were undergraduates. The number of responses was not proportional per bank as there were 11 respondents from FDH Bank, 11 from SB, 33 from NBS Bank, 88 from Standard Bank and 209 from NB and neither this is an indication of the banks market penetration.

Regression analysis of ATM Attributes and Customer Satisfaction with ATM Banking.

Model Summary									
Model	**R**	**R Square**	**Adjusted R Square**	**Std. Error of the Estimate**	**Change Statistics**				
					R Square Change	**F Change**	**df1**	**df2**	**Sig. F Change**
1	1.000a	1.000	1.000	.000	1.000	5.706E15	23	306	.000

a. Predictors: (Constant), ATM Fees charged, Appearance of corporate branding on ATM, Employee Friendliness, Issuing of Clean Notes, Waiting times on ATM, Quick replacement of lost ATM cards, Security and advise on ATM usage, Cash availability in ATMs, Service range at ATMs, Easy application process for ATM cards, User friendliness of ATM systems, Security at ATM, Readable ATM Slips, Convenient Location, Employee access, Fast return of swallowed ATM cards, Cleanliness of

ATMs, Accuracy of ATM transactions, Employee Speed, Privacy at ATMs, Easy access to ATM, Speed of ATMs, Employee effectiveness.

It is important to note first that the attributes that were adapted to measure customer satisfaction with ATM banking provided a perfect model (Table) where the R square of 1.000 denotes that all the attributes adopted provided 100 percent predictive capability of satisfaction with ATM banking with statistical significance $p<0.001$.

	Satisfaction Ratings				
Bank	**2**	**3**	**4**	**5**	**Total**
FDH Bank	0	0	11	0	11
MSB	0	0	11	0	11
NBM	77	33	55	44	209
NBS	33	0	0	0	33
Standard Bank	11	22	55	0	88
TOTAL					352

Table: Customer satisfaction with ATM banking

	Satisfaction Ratings				
Bank	**2**	**3**	**4**	**5**	**Total**
FDH Bank	0	11	0	0	11
MSB	11	0	0	0	11
NBM	77	33	77	22	209
NBS	11	11	11	0	33
Standard Bank	22	33	33	0	88
TOTAL					352

Table. Customer satisfaction with Bank

Table shows frequencies of respondents rating of their satisfaction with ATM banking and Table shows frequencies of respondents rating of their overall satisfaction with their respective banks both on a Likert scale of 1 to 5 where 1 was equal to 'very poor', 'very dissatisfied'; 2 was equal to 'poor', 'dissatisfied'; 3 was equal to 'neither poor nor good', 'neither dissatisfied not satisfied'; 4 was equal to 'good', 'satisfied' and 5 was equal to 'very good', 'very satisfied'.

Although all respondents from FDH Bank were satisfied with the performance of the ATM banking from their bank, but all the respondents were neither dissatisfied nor satisfied with their bank overall. The respondents from MSB were satisfied with the ATM banking from their bank and all of them were dissatisfied with their bank overall. NBM had 47 percent of its respondents satisfied with its ATM banking, 16 percent neither dissatisfied nor satisfied and 37 percent dissatisfied with its ATM banking. Equally the same percentages of respondents from NBM were overall dissatisfied and satisfied with the NBM. For NBS all respondents were not satisfied with its ATM banking, however 33 percent of the respondents were overall satisfied with their bank. For Standard Bank 55 respondents were satisfied with their bank's ATM banking, 22 were neither dissatisfied nor satisfied and 11 were dissatisfied. 33 respondents were overall satisfied with their bank, 33 were neither dissatisfied nor satisfied and 22 were dissatisfied with their bank.

The study results are interesting because although most respondents from FDH, MSB and Standard Bank were satisfied with ATM banking offered by their respective banks, they were not equally satisfied with their bank. Whereas all the respondents from NBS were not satisfied with the ATM banking from their bank, 33 percent of the respondents were still satisfied with their bank. It is a very unusual case with NBM where the same frequencies of responses apply to both satisfaction with ATM banking and the bank.

The data was therefore subjected to linear regression analysis to assess the predictive power of satisfaction with ATM banking performance on the satisfaction with the bank. Table 4 is the model summary of the regression analysis of satisfaction with ATM banking (independent variable) and satisfaction with bank (dependent variable). The adjusted R square of 0.403 denotes that satisfaction with ATM banking explains 40 percent of customers' satisfaction with their respective banks with statistical significance $p<0.001$.

This result demonstrates that ATM banking performance is one of the key elements that determine satisfaction with banks. ATM banking is the second popular access channel to banking products/services behind branch banking and this study further confirms the importance of ATM banking in determining customer satisfaction with banks.

Based on these results, banks need to consider ATM banking a significant area of customer service provision in the quest to achieve greater customer satisfaction with the bank itself in order to expand the base of satisfied customers. There are however other elements in the customer satisfaction equation with the bank that make up the other 60 percent which require further studies so that adequate knowledge is generated on factors that contribute to 100 percent customer satisfaction with banks. The study findings nonetheless support the research hypothesis that ATM banking performance has a positive and statistically significant relationship with customer satisfaction with banks.

To condense further the importance of ATM banking in the customer satisfaction equation, respondents were asked if they would consider switching banks because of poor ATM banking performance. 75 percent of the respondents indicated that they are unlikely to change banks because of poor ATM banking service against 19.1 percent who were certain about changing banks because of poor ATM banking service. Respondents were further asked whether they would recommend their bank irrespective of poor or good performance of ATM banking. 81.3 percent of the respondents indicated that they were likely to recommend their respective banks irrespective of the performance of ATM banking.

These responses point to the fact that much as ATM banking is key in providing customer satisfaction with banks, it is not adequate to lure customers from other banks since customers cannot switch banks due to poor ATM banking performance. Customers would not be attracted to join a bank because of newer efficient ATM technologies. ATM banking is not adequate for attracting new customers from other banks although it plays a significant role in determining customer satisfaction with the banks.

As noted that differences in education, culture, experience, and tolerance levels of service quality affect participants' perceptions of satisfaction in particular contexts, it would be interesting based on these findings to replicate this study in different contexts to further understand the effects of satisfaction with ATM banking performance on customer satisfaction with bank. It would also be necessary to investigate the other key elements that affect customer satisfaction with banks apart from ATM banking focusing on factors that would attract customers to switch banks as that knowledge would be ideal in developing competitive strategies to achieve larger customer bases and profitability with banks.

8.5 ATM use satisfying Customers

Earlier research by Brownlie has recommended that some consumers have positive attitudes towards ATMs based on dominant perceptions of convenience/ accessibility/ease of use. On the other hand, Reichheld and Sasser have recognized the benefits that customer satisfaction delivers to a bank. For instance, the longer a customer stays with a bank the more utility the customer generates. This is a result of a number of factors relating to the time the customer spends with a bank. Without usage of technology the banking sector cannot provide customers with a satisfactory service. Effective service delivery is a new or significantly improved service concept that is taken into practice.

According to, Patricio et al. customers will use different service delivery systems dependent on their assessment of each channel and how it contributes to the overall service offering. Hence service satisfaction will not merely be based on isolated service encounters and experiences but rather on the overall feelings of satisfaction. With automated teller machines networks already in place in most of the urban areas, the drive is now focused towards the rural areas where the use of automated teller machines is still uncommon.

It can be observed that the operation and use of ATM services in the financial sector, has contributed a lot in changing the way in which financial services and products are being delivered to the banks clients. As the say goes that, for every step forward (development), a lot of challenges must have been faced, fought and overcome. Thus the development of ATM saw the emergency of some challenges for the industry as customers keep demanding for better service, while the financial institutions are very busy searching for the most efficient way by which they can improve on their service provision.

Implementation of improved customer interaction initiatives and one-to-one services delivery becomes too expensive if these changes are done on a channel-specific basis. Many banks today are implementing channel integration and Customer Relationship Management (CRM) strategies around which improved solutions for ATMs and other customer service channels can be built.

According to the IDC study "Web-Enabled ATMs: Implications for Online Banking and Multichannel Delivery," banks that are embarking on a channel integration strategy may find integrating online, kiosk and ATM solutions to be a valid first step. CRM enabling of the result would then form a basis for the bank's remaining customer service channels.

Applications included in online banking solutions can have appropriate cross links into the ATM channel. Electronic bill presentment and payment (EBPP), account aggregation and online lending services are commonly configured and utilized in an online environment. In such cases, quick checks or adjustments to the configuration of these services can then be offered at ATMs.

Increasing the number of customer connections in this manner not only increases the opportunities for recognizing and delivering revenue generating services, but it also benefits retention of high value customers. A recent Celent Communications report revealed that offering online banking solutions improved customer retention. According to that report, "banks are discovering that active users of their online banking services are much less likely to jump ship to another institution," and that such services increase the value of active customers to the bank by twenty-five percent. Integrating online and ATM services can extend that stickiness to the ATM, resulting in an even more powerful platform for customer focused services.

Integration of existing service channels, as well as potential channels such as mobile phones and PDAs, allows driving of content appropriate to each channel and the customer using that channel device in real time. Such integration leads to more opportunities for expanding the transaction set and related revenue streams.

8.5.1 Expanded Transactions

- ATMs continue to increase their transaction capabilities. There are increasing options available for automating the handling of more complex transactions, such as check cashing, bill presentment, money order issuance, ticket printing and bulk cash recycling. These capabilities expand the suite of value delivery services that ATMs can deliver.

- The existence of convenience fees for non-customer use of a bank's ATMs shows that consumers are more than willing to pay for convenience. Consumers have come to recognize the benefits of having desired transaction capabilities close at hand and are willing to pay for this convenience. Thus, there is a relationship between the increasing capabilities of ATMs to fulfill convenient transactions and the revenue potential of such systems.
- The fact is that the unique aspects of ATMs are under-utilized among the services and functions deployed within networks today. Functions such as statements on demand, ticket dispensing or ticket printing, check cashing and various location-based services all can be aimed at a convenience orientation and be chargeable by event.

8.5.2 Location Based Services

- An ATM has another distinct advantage over other channels of the bank. Wherever a consumer goes, there is an ATM nearby whose location is known. When a consumer utilizes a particular ATM, the network can then know where that consumer is. A merging of location, consumer information and banking relationships opens the way to targeted commerce options at every ATM.
- An example of this is an ATM whose location is near a movie theatre. Based on a selection profile, the consumer may be offered (during the transaction authorization period) a short preview of a film and the option to reserve tickets for a showing. The bank would charge a convenience fee for this service.
- Or consider a businessperson traveling out of town. Based on this fact, the time of day, and the expected traffic at the ATM, a list of neighborhood restaurant recommendations may be shown. These placements may be a product sold by the bank to neighborhood businesses.
- Thus, customer focused services at an ATM not only build relationships with the bank's retail customers, but also serves as a vehicle for building content and transaction relationships with the bank's commercial customers. Such a platform allows for customization of solutions that are specific to individual banks' business needs and relationships.

8.6 Problems with ATM

8.6.1.Problem faced by customers

a. Sometimes, customers are facing problem with ATM when their account is debited. They didn't receive cash while the account is debited.

b. The most common problem known to all when one is new and do not know how to insert the ATM card. They wrongly insert the ATM card and This problem is more common with new ATM users who are not familiar with ATM machine. This same incident had happened with me when I am new with my ATM card.

c. Sometimes ATM machine does really bad with the customers, as when one didn't able to receive slip of account balance, which happens when ATM machine get older or get some serious technical problem.

Money transfer problem is another big problem

It generally happens for four reasons.

a. Payment Declined - By Your Bank

The most common reason for an unsuccessful card payment is that your bank has declined it. This means they have sent a 'Do Not Honour' response to Transfer Wise. It's normal for them to stop a payment that they might perceive to be unusually large or to a new and unknown recipient - just in case.

b. Browser Issues

Some anti-virus or firewall software can affect your card payment. With anti-virus software, please add Transfer Wise to the list of "safe merchants". Please also make sure that any anti-virus or firewall software that you have installed will allow pop ups.

c. Insufficient Funds

Your bank might also stop your card payment if there isn't enough money in your account or if you have a per-transaction limit. They could also be treating your payment as 'cash withdrawal'. The card limits vary from card to card and from bank to bank

Machine don't accept the ATM cards

Some time machine don't accept the ATM cards.A variety of factors can disrupt this process and result in the card being declined.

a. Inadequate Balance

A common reason for a card being declined is a balance that is too low to support the requested transaction. You may have less money in your account than you thought, or a previously-made transaction, like a deposit, may not have cleared. Take into consideration both your request and the associated ATM fee.

b. Account Changes

If your account is frozen for any reason, your ATM card will likely be declined. For example, if you recently experienced a number of overdrafts, reported a card missing or stolen, or had a fraud alert put on your card, a machine will likely decline it and, in some cases, even retain the card and not give it back to you. If your card has expired, this, too can trigger a decline notice. Many financial institutions also limit how much money you can take from an ATM in a 24-hour period. If you've surpassed this number, your card may be declined.

c. Wrong PIN

If you enter the wrong PIN, or personal identification number, your card may be declined. In most instances, the machine will prompt you to retry the PIN, but if you enter it wrong repeatedly, the ATM may hold your card in an effort to ensure that it is not being used illegally by someone else.

4. Technical Glitches

Sometimes ATMs are undergoing service or experiencing a computer malfunction that results in declined cards. ATMs can also be emptied of cash, especially on busy weekends. While most low or empty machines automatically post an "unavailable" icon of some sort to alert potential users to the problem, you may also experience a straight decline instead.

8.6.2 Problems faced by BANK

a. Theft money from ATM's

Bank thefts are getting sophisticated by the day. Let alone retrieving your money, you can't even get to know how it was stolen in many cases. Some customers even lost lots of amount after using that ATM.

b. Cloning problem in ATM

A type of fraud which occurs when an ATM is compromised by a skimming device, a card reader which can be disguised to look like a part of the machine. The card reader saves the users' card number and pin code, which is then replicated into a counterfeit copy for theft. Cloning occurs mainly by replicas presented in ATM system. It will extract the data presented in the magnetic tape of a card .The data is send to the memory chip which is attached to the replica The closed circuit camera will extract the password which we are entered in our ATM system. They will prepare fake ATM's cards and draw the money from those cards.

c. Problems of sms alert of atm's

If server gets down, such kind of problem occurs in ATM system. You will be very disappointed when you didn't get any sms regarding whether your money debited or credited.

d. Problem of Skimmers

Credit card skimming is a type of credit card theft where crooks use a small device to steal credit card information in an otherwise legitimate credit or debit card transaction. Credit card skimmers are often placed over the card swipe mechanism on ATMs and gas stations, but they skimmers can be placed over almost any type of credit card reader. With ATMs, the crooks may also place a small, undetectable camera nearby to record you entering your PIN. This gives the thief all the information needed to make fake cards and withdraw cash from the cardholder's checking account.

8.7 ATM Banking Attributes

Automated Teller Machines (ATM) are devices used by bank customers to process account transactions. Typically, a user inserts into the ATM a special plastic card that is encoded with information on a magnetic strip. The strip contains an identification code that is transmitted to the bank's central computer by modem. To prevent unauthorized transactions, a personal identification number (PIN) must also be used by the user using a keypad. The computer then permits the ATM to complete the transaction; most machines can dispense cash, accept deposits, transfer funds, and provide information on account balances. Banks have formed cooperative, nationwide networks so that a customer of one bank can use an ATM of another for cash access, by extension all commercial bank's ATM in Nigeria are inter-connected.

Globally, Automated Teller Machines (ATMs) have been adopted and are still being adopted by banks. They offer considerable benefits to both banks and their depositors. The machines can enable depositors to withdraw cash at more convenient times and places than during banking hours at branches. In addition, by automating services that were previously completed manually, ATMs reduce the costs of servicing some depositors of demand. These potential benefits are multiplied when banks share their ATMs, allowing depositor of other banks access their account through a bank's ATM.

Banks have become the principal deployers of ATMs. Two reason for this are that they want to increase their market share, although due to the prevalence of ATMs, it is not likely to be the primary means by which ATMs increase profitability for most banks, or/and above a certain level of operations, the cost of a single transaction performed at an ATM is potentially less than the cost of a transaction conducted from a teller, as ATMs are capable of handling more transactions per unit of time than are tellers.

In Nigeria the deployment of ATM by banks and its use by bank customers is just gaining ground and has burgeoned in recent times. This has happened especially after the recent consolidation of banks, which has in all probability, made it possible for more banks to afford to deploy ATMS or at least become part of shared networks. The increased deployment of ATMs in the banking sector has made the issue of technology relevance important.ATM services have a history that is less than ten years. At first, they were operated as elitist services designed for those desirous of exclusive services. Cards were rare and the process for obtaining them tortuous.

Presently, the use of ATM cards has been widely promoted. Banks no longer appear to want personal contact with their customers. Some banks have resorted to penalizing the customer as it were, for not possessing an ATM card, by debiting the account of such a customer for withdrawing below a certain amount across the counters. As reported, although only a bank had an ATM in 1998, by 2004, fourteen of them had acquired the technology.

As discovered that the adoption of ICT in banks has produced largely positive outcomes such as improved customer services, more accurate records, ensuring convenience in business time, prompt and fair attention, and faster services etc. Also, the banks' image is improved creating a more competent market. Work has also been made easier, and more interesting, the competitive edge of banks, relationship with customers, and the solution of basic operational and planning problem has been improved. The debit card transaction rose by 93 percent over pervious years owing to aggressive roll out initiatives by Nigerian banks, powered by interswitch network the number of ATM transactions through inter switch network had increased from, 1,065,972 in 2004, to 21,448,615 between January 2005 to March 2012.

This is a rise of 92.6 percent with respect to the pervious years. More than 1700 ATMs have been deployed on the network, while about 12 million cards have been issued by 18 banks as at March 2012.

A recent survey conducted by Intermarc Consulting Limited revealed that ATM services provided by Nigeria by banks and non-financial institutions stood as the most popular e-business platforms. The report showed that awareness for various banking services by banks is mostly limited to the traditional banking services. The findings shows that 99% of the respondents were aware of savings account, while 92 where aware of current accounts and 72 percent are aware of local money transfer services. However, among the more modern banking services such as electronic banking, internet banking, point of sales (POS) transactions, money transfer, ATMS emerged as the most popular with 96 percent awareness level ATM awareness also ranked higher than awareness level about current accounts and slightly below savings account.

Marketing literature defines a product/service as bundle of features (attributes) that bears ability to satisfy implicit or explicit needs/wants of customers. Service quality is therefore a measure of how well a product/service is delivered to meet the expectations of customers. Literature thus asserts service quality as an antecedent to customer satisfaction.

Similarly, ATM banking is a product/service that has attributes/features that meet implicit and/or explicit needs of bank customers. To achieve customer satisfaction the attributes have to deliver service quality to customers' expected threshold. Therefore before customer satisfaction is delivered through ATM banking, managers ought to understand the features/attributes of ATM banking that provide service quality because these are the factors that influence performance superiority to influence customer satisfaction.

Various empirical studies identified ATM banking attributes that were adopted by this study. These attributes include number of ATMs per ATM station; issuance of clean notes; readable slips; accessibility to a wide range of banking services via ATMs, appearance of corporate branding on ATMs, cleanliness of ATMs and ATM stations; ease of application process for ATM cards; privacy when using ATMs; advice on ATM usage and ATM security; employee accessibility to solve ATM problems; convenient location of ATMs; ATM fees; security at ATM stations; bank employee friendliness when assisting on ATM issues; quick replacement of lost cards; fast return of swallowed cards; employee speed in responding to ATM problems; employee effectiveness in solving ATM problems; waiting times at ATMs; ATMs not out of order; accuracy of ATM transactions; cash availability in ATMs; speed of ATM transactions and user friendliness of ATM systems.

There were a total of 25 ATM attributes adopted from the empirical studies. These ATM banking attributes fall within the five dimensions of service quality that include: tangibles, reliability, responsiveness, assurance and empathy. Tangibles are the physical facilities, equipment as well as appearance of personnel; reliability is the ability to perform the promised service dependably and accurately; responsiveness is willingness to help customers and offer prompt service; assurance is knowledge and courtesy of employees and their ability to inspire trust and confidence; and empathy is caring, individualized attention the firm provides its customers.

The study adopted 25 attributes of ATM banking from empirical studies falling into the five dimensions of service quality for the analysis of the factors that affect customers' satisfaction with ATM banking and were operationalised through a multiple scale performance measurement instrument.

Chapter 9
Repair and servicing of Automated Teller Machine (ATM)

9.1 Introduction

TMs are also widespread, allowing customers to carry out bank transactions without the assistance of a teller. ATMs now provide a growing variety of other services, including stamp, phone card, and ticket sales. ATMs are used throughout the world to extend financial services to customers of financial institutions and provide customers with variety of secure financial services without the assistance of human teller which includes depositing, withdrawing, or transferring funds, exchanging currency, selecting and viewing account balances, statements, or credit limits, purchasing items such as stamps or gift cards, managing accounts and profiles, etc.

As with any mechanical device, ATMs are prone to malfunction. With variety of diagnostic and repair tools to complete their work, ATM technicians diagnose issues with malfunctioning ATMs and work to fix the equipment to ensure proper function. When an ATM malfunctions, the ATM is often first seen by a low-level technician, who possesses a basic understanding of the ATM. Level one technicians may be an owner or custodian of an ATM, tellers, retail bankers, or other employees working at a bank having a minimal knowledge level or basic training regarding ATM repair.

If level one technician is unable to fix the problem, a level two technician, who may have a medium level of training and understanding of the ATM, may be beckoned to fix the problem. A level three technician, who may have a higher level of training and understanding, may ultimately be required to fix the ATM's malfunction. The cost of ATM repair may increase with the technical level of the technician repairing the machine. Also, the amount of time required to repair the ATM may also increase with the technical level of the technician since higher level technicians must travel to the site of the malfunctioning ATM than higher level technicians. To provide a high level of customer service, financial institutions and ATM owners and operators may seek to minimize the down time of ATM due to technical malfunctions or problems. Thus, enhanced systems and methods of quickly and efficiently repairing ATMs are desired."

When equipment breaks down, many repairers travel to customers' workplaces or other locations to make the necessary repairs. These workers, known as field technicians, often have assigned areas where they perform preventive maintenance on a regular basis. Bench technicians work in repair shops located in stores, factories, or service centers. In small companies, repairers may work in both repair shops and at customer locations.

When ATMs malfunction, computer networks recognize the problem and alert repairers. Common problems include worn magnetic heads on card readers, preventing the equipment from recognizing customer bankcards; and "pick failures," preventing the equipment from dispensing the correct amount of cash. Field technicians travel to the locations of ATMs and usually repair equipment by replacing defective components. Components that cannot be replaced are brought to a repair shop where bench technicians perform the necessary repairs. Field technicians perform routine maintenance on a regular basis, replacing worn parts and running diagnostic tests to insure that the equipment functions properly.

ATM include a damaged card reader, which prevents computers from reading a customer's card. This problem is diagnosed by a computer and then a technician is notified to fix the problem. ATM technicians usually remove damaged parts, install new parts, and take the damaged parts to a shop to be fixed. Technicians also perform preventative maintenance on ATM machines.

Automated teller machine repair specialists fix and install parts in automated teller machines (ATMs). ATM machines is where cash is withdrawn. Some ATMs sell other products such as stamps and phone cards.

Fig technician analyzing the fault

Most technicians perform repairs in clean, brightly lit environments. Likewise, machines are usually housed in ventilated, climate controlled buildings. Technicians often perform onsite repairs, so they travel frequently. Technicians repairing ATM machines sometimes work in cramped spaces.

Since office equipment and ATMs are vital for business operations, a technician's services are usually available 24 hours a day. Technicians often work nights, weekends, and holidays, but those with seniority can choose their shifts. Those repairing cash registers or non vital office equipment usually work 40 hour weeks. Over 12 percent of technicians work 50 or more hours a week.

9.2 Repairing work

Repairers usually work in clean, well-lighted surroundings. Because computers and office machines are sensitive to extreme temperatures and to humidity, repair shops are usually air-conditioned and well ventilated. Field repairers must travel frequently to various locations to install, maintain, or repair customer equipment. ATM repairers may have to perform their jobs in small, confined spaces, which house the equipment.

Because ATMs are critical for many organizations to function efficiently, data processing repairers often work around the clock. Their schedules may include evening, weekend, and holiday shifts; shifts may be assigned on the basis of seniority. Office machine repairers usually work regular business hours, because the equipment they repair is not as critical.

9.3 Parts of ATM which can be repaired

There are many items on an ATM that can be repaired as opposed to being replaced, and this is usually a more cost-effective alternative. Some of the key modules that are the most popular items for repair include, cash dispensing mechanism, receipt printer, power supply, card reader, display, keypad and electronic circuit boards. The cash dispensing mechanism is one of the most popular items sent in for repair since it has to hold and dispense out the currency. This technology typically relies on belts and rubber wheels to pass the currency out to the customer.

ATM servicing/repair is much more than just the replacement of belts and gears. As with any component, there are a number of adjustments that must be made once a repair has been completed. Whether it is delivering currency, providing a paper receipt, or even adjusting the performance of an electronic circuit board, tuning a component to ensure maximum performance is a key factor in a successful repair.

It is critically important that an ATM field technician or bench technician have the proper tools and ATM training to provide the best possible service. Knowing what updates are available/necessary, having certified repair parts, having a dedicated ATM training system along with competent ATM technicians will go a long way in providing qualified ATM component repairs.

Receipt Tape

Other than the cash inside the machine, receipt tape will be the most common part of ATM that you need to replace. The good news is that this isn't all that hard to do, nor is the tape particularly expensive. If you hire a technician to do regular maintenance, changing the receipt tape will likely be part of the routine. Otherwise, you'll have to keep an eye on it and take care of it yourself.

Rollers

ATMs have rollers that help to receive and dispense user's ATM cards, as well as dispense cash from the machine. If they start to wear down, they'll need to be replaced, so checking on them during regular maintenance is important. It also wouldn't hurt to clean these rollers every once in awhile so they work properly without causing your customers any problems.

Screen

Odds are, you won't have to replace your ATM's screen too often. However, with most ATMs having touch screens, it would be wise to learn it regularly. Take a second to think about the volume of people using your ATM and you'll realize how important it is to clean the screen. Also, during routine maintenance, you should have the technician calibrate the screen so that everything lines up accordingly, preventing users from unknowingly choosing the wrong option when they touch the screen.

Fig ATM screen

Card Reader

The card reader slot is one part of the ATM that has to remain in working order. Without it, the machine wouldn't be able to recognize any of the numbers on a user's ATM card, rendering the entire machine useless. If you own an ATM, it wouldn't hurt to have a backup card reader on hand in case you need to install a new one right away. Without a functioning card reader, your ATM can't process anything, so you can't afford to be without one for long.

Fig card reader

Cash Dispenser

Behind the card reader, the cash dispenser may be the second most important part of an ATM. If there's a problem with it, people may receive too little or too much money from the machine; either way, that's a big problem. You may want spare parts, including the rollers mentioned above, that make up the cash dispenser available in case something needs to be replaced.

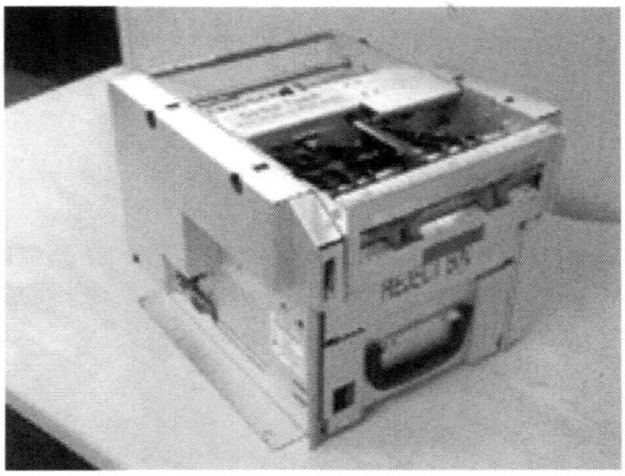

Fig cash dispenser

Glossary

ACD/ATM application: Application software which directly or indirectly controls the operation of the ACD/ATM. It is normally developed by the ACD/ATM manufacturer, but may also be developed by outside companies or directly by banking establishments. It may be installed on the ACD/ATM by maintenance staff (diskette) or by the download server.

ACD/ATM controller (AAC): Placed under the responsibility of the purchaser, this handles two main functions: control of the ACD/ATM and bank management (authorization, transaction reports, relations with the issuer, etc).

Bank teller: More accurately, this is the ACD/ATM "operator", responsible for "account" management (account balancing, replenishment of banknotes, repossession of seized cards, etc) and day-to-day operations (replacement of supplies, etc). Two very differerent roles need to be distinguished:

- the bank teller responsible for banking operations,
- the maintenance engineer responsible for maintenance.

Black box: This is a secure enclosure for safeguarding secrets and protecting sensitive information (personal identification number and encryption keys), in particular information interchanged between the ACD/ATM and the AAC (ACD/ATM controller).

Card: Card enabling the target of evaluation to identify and then authenticate the cardholder.

Cardholder: The holder of a bank card issued by an issuer.

Cash dispenser: This is a module used to extract banknotes from their cassettes and deliver them to the customer (and "swallow" them if left by the cardholder).

Device drivers: These are programs (often in the form of an API and sometimes included in the ACD/ATM application) for controlling the peripheral devices (floppy drives, printers, etc). They may be installed on the ACD/ATM by the maintenance engineer (diskette) or by the download server.

Download server: Normally the responsibility of the purchaser, this is used to install new applications, new device drivers or any other software (firmware, etc) on the ACD/ATM. Other servers (or, at least, other functions) must also be taken into account: remote monitoring server, etc.

Firmware: This is "software" which is built into the peripheral devices. It is normally installed by the manufacturer and updated by the maintenance engineer (PROM), but may sometimes be downloaded.

Issuer: This is the banking organization that supplies the card to the cardholder.

Keypad: This is a device used by the cardholder to enter his or her personal identification number and select an operation, an account, an amount, etc or enter the amount, account number, telephone number, etc. The keypad and the black box may be combined within the same device.

Maintenance engineer: Broadly speaking this is the person responsible for ensuring that the ACD/ATM is available (repairs, servicing, software updates, replacing of components, etc). The maintenance engineer may need to use the machine's floppy drive to access particular items (statistics files, etc) or to install software.

Manufacturer: designs and builds the ACD/ATM, often by assembling elements (peripherals, microcomputer) sourced from various suppliers. The manufacturer also designs and maintains the device driver, firmware, etc and often designs and maintains the ACD/ATM software.

Operating system: This is normally a standard off-the-shelf system

Operator: A person responsible for maintaining, replenishing, etc the ACD/ATM. Two very different roles need to be distinguished:

- the maintenance engineer responsible for maintenance,
- the bank teller responsible for banking transactions.

Parameter file: A set of files used by the purchasing bank to customize the operation of the ACD/ATM on the basis of ACD/ATM software (operations available, chain of events, advertising screens, etc). Customization is the responsibility of the purchasing bank which sometimes delegates the task to different partners, particularly the manufacturer.

Customization can be updated via a diskette, by remote parameter setting (AAC) or by downloading (download server).

Purchaser: The organization with responsibility for the ACD/ATM and for the fiduciary currency placed in the ACD/ATM, for the ACD/ATM controller and for its operation. The purchaser often develops the ACD/ATM application and normally manages the parameter files (customization).

Reader / Reader interface controller: Peripheral device which can read a card (various kinds of technology: B0' chip, EMV, ISO2 stripe), write information (chip), disable an application or the chip, submit the personal identification number to the chip and optionally seize the card.

Telecommunication drivers: Generally added to a telecommunication card, these enable the ACD/ATM to communicate with the AAC (ACD/ATM controller), and, where appropriate, with other servers (maintenance, downloading, etc). ACDs/ATMs normally communicate in X25 mode via leased lines etc, but other solutions will probably soon emerge (TCP/IP, Internet, etc).

ATM Card:- Debit card use by banks customer in making transactions via ATM. The card is a complex circuit that process microprocessors with single chips that contain the complex Arithmetic and logic unit of computers. It provides access to customers to perform balance inquiry, mini statement and cash withdrawal as well as transfers through the use of Automated teller Machines. This Debit card can also be used for internet online and POS transactions.

Chip card:- This is a card containing one or more computer chips or integrated circuits for identification, data storage or special purpose processing used to validate personal identification numbers, authorize purchases, verify account balances and store personal records.

Electronic Data Interchange (EDI): The transfer of information between organization machines readable form.

Electronic money:- Monetary value measured in currency units stored in electronic form on electronic device in the consumer's possession. This electronic value can be purchase and held on the device until reduced through purchase or transfer.

Mobile Banking:- This is a product that enables the bank to offer customers to access services anywhere. Customer can make their transactions anywhere such as account balance, transaction enquires, stop checks, and other customers services instructions balance inquiry, account verification, bill payment, electronic fund transfer, account balances, updates and history, customer service via mobile, transfer between account etc.

Banking: The business receiving money from outside as deposit for safe keeping and making payment accrued which the money is due for payment and also advancing loans to those who keeps money for safe keeping as well as the public at large.

Account: Is a period during which transactions takes at the end and in which settlement must be made

Banker: These refers to group of persons who receive money from individuals for safe keeping in the agreement that he will refund the said amount collected either on demand or at some certain date agreed upon.

Cheque: It is an order written by the owner to the banker to pay on demand to the bearer of the cheque.

Dishonored Cheque: A cheque, which the bank refused to attend to for one reason or the other e.g irregular signature, post dated cheque etc.

Loan: The amount of money that is lend out and that must be repaid within an agreed rate of interest usually for specific period of time.

Pass Book: It supplied to the customer by the bank in which all entries are made of all deposits and withdrawals that are made.

Unpresented Cheque: These are cheque drawn by the drawer but not yet presented for payment by the bearer.

Overdraft: Is usually created on a current account unlike a loan account where only periodical payments are made. An overdraw account is a running account where drawing and deposits are made. An overdraft account is a running account where drawing and deposits are made. In an overdraft account as frequently as may be needed or received in connection with the business to meet the account related.

Payment System – A financial system that establishes that means for transferring money between suppliers and of fund, usually by exchanging debits or Credits between financial institutions.

Point Of Sale (POS) Machine - A Point-of-Sale machine is the payment device that allows credit/debit cardholders make payments at sales/purchase outlets. It allowed customers to perform the following services Retail Payments, Cashless

Payments, Cash Back Balance Inquiry, Airtime Vending, Loyalty Redemption, Printing ministatement etc.

Smart Card – A Card with a computer chip embedded, on which financial health, educational, and security information can be stored and processed.

Transaction Alert - Our customers carry out debit/credit transactions on their accounts and the need to keep track of these transactions prompted the creation of the alert system by the Bank to notify customers of those transactions. The alert system also serves as notification system to reach out to customers when necessary information need to be communicated.

Western Union Money Transfer (WUMT) - Western union Money transfer is a product that allowed people with relatives in Diaspora who may be remitting money home for family up-keep, Project financing, School fees etc. Nigerian Communities known for having their siblings gainfully employed in other parts of the world are idle markets

ICT: Acronym for Information and communication technology.

ATM: Acronym for Automated Teller Machine

MIS: Acronym for management information system

Online-real-time: Electronic banking through internet and computer network

References

Adams, A. S. and Thieben, K. A. (1991), Automatic teller machines and the older population. Applied Ergonomics, 22, 85 -90.

Aldlaigan, A., and Buttle, A. (2002). SYSTRA-SQ: a new measure of banks service quality. International Journal of Service Industry Management. 13(4), pp. 362–381.

Al-Hawari, M. and Ward, T. (2006).The impact of automated service quality on financial performance and the mediating role of customer retention. Journal of Financial ServiceMarketing.10(3), pp.228-43.

Andaleeb, S.S. and Basu, A. K. (1994). Technical complexity and consumer knowledge as moderators of service quality evaluations in the automobile service industry. Journal of Retailing Vol.70(4) pp367–379.

Anderson, E., & Sullivan, M. (1993). The antecedents and consequences of customer's satisfaction for firms. Marketing Science, 12(2), 125-143.

Anderson, E.W. and Fornell, C. (1994).A customer satisfaction research prospectus. Services marketing: new directions in theory and practice.ThousandOaks, CA: Sage Publications, pp. 241–68.

Asubonteng, P., McCleary, K.J. and Swan, J.E. (1996). SERVQUAL revisited: a criticalreview of service quality. Journal of Service Marketing10(6):62– 81.

Athanassopoulos, A. (2000).Customer satisfaction cues to support market segmentation andexplain switching behaviour. Journal of Business Research.47(3),pp.191-207.

Babakus, E. and Boller, G.W. (1992), "An empirical assessment of the SERVQUAL scale", Journal of Business Research, Vol. 26 No. 6, pp. 253-268.

Banker, R.D. and Kauffman, R.J. (1988).Strategic contributions of information technology: An empirical study of ATM networks, proceedings of the ninth International Conference for Information Systems, Minneapolis, MW.

Berkley, B.J. and Gupta, A. (1994). Improving service quality with information technology, International Journal of Information Management, Vol. 14, pp. 109-21.

Bolton, R.N. and Drew, J.H. (1991). A multistage model of customers' assessment of service quality and value, Journal of Consumer Research, Vol. 17, March, pp. 375-84.

Boulding, W., Kalra, A., Staelin, R. and Zeithmal, V.A. (1993). A dynamic process model of service quality: from expectations to behavioural intentions. Journal of Marketing Research Vol.30 pp 7–27.

Brogowicz, A.A., Delene, L.M. and Lyth, D.M. (1990).A synthesized service quality model with managerial implications.International Journal of Service Industry Management. 1(1), pp. 27-46.

Brown, S.W. and Swarth, T.A. (1989). A gap analysis of professional service quality. Journal of Marketing Vol. 53, pp 92–98.

Cadotte, E.R., Woodruff, R.B. and Jenkins, R.L. (1987).Expectations and norms in models of consumer satisfaction.Journal of Marketing Research.24 (August): pp. 305-314.

Carman, J.M. (1990). Consumer Perception of Service Quality: An Assessment of the SERVQUAL Dimensions.Journal of Retailing.66, pp.33-55.

Churchill, G.A. and Surprenant, C. (1982).An investigation into the determinants of customer satisfaction.Journal of Marketing Research, Vol. 19, pp. 491-504.

Cronin, J.J. and Taylor, S.A. (1992).Measuring service quality: A reexamination and extension.Journal of Marketing, 56, July, pp. 55-68.

Dabholkar, P.A. (1996). Consumer evaluations of new technology-based self-service operations: an investigation of alternative models, International Journal of Research in Marketing, Vol. 13 No. 1, pp. 29-51.

Dabholkar, P.A., Shepherd, C.D. and Thorpe, D.I. (2000), "A comprehensive framework for service quality: an investigation of critical conceptual and measurement issues through a longitudinal study", Journal of Retailing, Vol. 76 No. 2, pp. 131-139.

Davies, F., Moutinho, L. and Curry, B. (1996). ATM users attitudes: A neural network analysis.Marketing intelligence & Planning, 14(2), 26-32.

Day, R.L. (1984). Modeling choices among alternative responses to dissatisfaction. In Advances in Consumer Research 11th Ed. William D. Perreault. Atlanta, GA: Association for Consumer Research, pp. 496-499.

Dos Santos, B.L. and Peffers, K. (1993).The effects of early adoption of information technology: An empirical study.Journal of Information Technology Management. Vol. 5(1) 1-13.

Ennew C.T, Reed GV, and Binks M.R. (1993), "Importance — performance analysis and the measurement of service quality", European Journal of Marketing Vol. 27(2): pp.59–70.

Fick, G.R. and Ritchie, J.R.B. (1991). Measuring service quality in the travel and tourism Industry. Journal of Travel Research, Vol. 30 No. 2, pp. 2-9.

Fishbein, M. and Ajzen, I. (1975).Belief, attitude, intention and behavior: An introductionto theory and research. Addison-Wesley, Reading, Massachusetts.

Frost, F.A. and Kumar, M. (2000). INTSERVQUAL: an internal adaptation of the GAP model in a large service organization, Journal of Services Marketing, Vol. 14 No. 5, pp. 358-77.

Glaser, P.F. (1988). Using technology for competitive advantage: the ATM experience of Citicorp in managing innovation cases from services industries. Guide, B.R. and Quinn, J.B. (eds) National Academy Press, Washington DC pp 108–114.

Goode, M. and Moutinho, L. (1995).The effects of free banking on overall satisfaction: the use of automated teller machines.International Journal of Bank Marketing. Vol. 13(4) 168–190.

Gotlieb, J.B., Grewal, D. and Brown, S.W. (1994). Consumer satisfaction and perceived quality: complementary or divergent constructs? Journal of Applied Psychology, Vol. 79 No. 6, pp. 875-85.

Gronroos, C. (1984). A service quality model and its marketing implications. European Journal of Marketing.18(4), pp. 36-44.

Halstead, D., David H., and Sandra L.S. (1994).Multisource effects on the satisfaction formation process. Journal of the Academy of Marketing Science, 22 (Spring): 114-129.

Haywood-Farmer, J. (1988). A conceptual model of service quality. International Journal of Operations & Production Management.8(6), pp. 19-29.

Howcroft, J.B. (1991).Customer satisfaction in retail banking. Service Industry Journal, January, pp. 11-17.

Johnston, R. (1995). The zone of tolerance: exploring the relationship between service transactions and satisfaction with the overall service. International Journal of Service Industry Management, 6(2), pp. 46-61.

Joseph, M. and Stone, G. (2003).An empirical evaluation of US bank customer perceptions of the impact of technology on service delivery in the banking sector. International Journal of Retail & Distribution Management, 31(4),pp.190-202.

Kantrow, Y.D. (1989). ATMs called 'only hope' to minimize costs, cope with shrinking labour force. American Banker.

Khan, A.M. (2010) An empirical study of ATM service quality and customer satisfaction in Pakistani banks. European Journal of Social Sciences. Vol.13 (3) 333–344.

Kheng, L.L., Mahamad, O., Ramayah, T. (2010) The impact of service quality on customer loyalty: A study of banks in Penang, Malaysia. International Journal of Marketing Studies. Vol.2 (2) 57–66.

Komal, S.S. (2009) Impact of ATM on customer satisfaction (A comparative study of SBI, ICICI and HDFC banks).Business Intelligence JournalVol. 2(2) 276–287.

Laderman, F.S. (1990). The public policy implications of state laws pertaining to Automated Teller Machines. Economic Review (Federal Reserve Bank of San Francisco pp. 43–58.

Levesque, T.J., and McDougall, G.H.G. (1996).Determinants of customer satisfaction in retail banking. International Journal of Bank Marketing.14(7), pp. 12-20.

Lewis, R.C. and Booms, B.H. (1983).The marketing aspects of service quality. Emerging perspectives in service marketing, In Berry, L.L., Shostack, G. and Upah, G. (eds.) American Marketing Association. Chicago, IL pp. 99-107.

Magi, A. and Julander, C.R. (1996). Perceived service quality and customer satisfaction in a store performance framework: An empirical study of Swedish Grocery Retailers. Journal of Retailing and Consumer Services, Vol.3(1) pp 33–41.

Mattson, J. (1992). A service quality model based on ideal value standard, International Journal of Service Industry Management, Vol. 3 No. 3, pp. 18-33.

Moutinho, L. and Brownlie, D.T. (1989). Customer satisfaction with bank services: a multidimensional space analysis. International Journal of Bank Marketing, 7(5), pp. 23-27.

Mwatsika, C. (2014). Customers' satisfaction with ATM banking in Malawi. African Journal of Business Management.Vol.8 (7) 218-227

Newman, K. (2001). Interrogating SERVQUAL: A critical assessment of service quality measurement in a high street retail bank. International Journal of Bank Marketing, 19(3), pp. 126-139.

Nunnally, J.C. and Bernstein, I.H. (1994).Psychometric Theory, 3rd ed., McGraw-Hill, New York, NY.

Oh, H. (1999). Service quality, customer satisfaction and customer value: a holistic perspective", International Journal of Hospitality Management, Vol. 18, pp. 67-82.

Parasuraman, A., Zeithaml, V.A. and Berry, L.L. (1985).A conceptual model of service quality and implications for future research. Journal of Marketing.49, Fall, pp. 41-50.

Parasuraman, A., Zeithaml, V.A. and Berry, L.L. (1988). SERVQUAL: A multiple-item scale for measuring consumer perceptions of service quality. Journal of Retailing, Spring, pp. 12-40.

Patrı´cio, L., Fisk, R.P. and Cunha, J.F. (2003).Improving satisfaction with bank service offerings. Managing Service Quality, 13 (6), 471-482.

Philip, G. and Hazlett, S.A. (1997), "The measurement of service quality: a new P-C-P attributes model", International Journal of Quality & Reliability Management, Vol. 14 No. 3, pp. 260-286.

Rust, R.T. and Oliver, R.L. (1994). Service quality insights and managerial implications from the frontier in Roland T. Rust and Richard L. Oliver (eds), Service Quality: New Dimensions in Theory and Practice, Thousand Oaks, CA: Sage Publications pp 1 – 9.

Sachdev, S. B. and Verma, H.V. (2004).Relative importance of service quality dimensions: A multi-sectoral study. Journal of Services Research, Vol. 4(1).

Seth, N.S., Deshmukh, S.G., and Vrat, P. (2005). Service quality models: a review. International Journal of Quality & Reliability Management.22(9), pp. 913-949.

Spreng, R.A. and Mackoy, R.D. (1996). An empirical examination of a model of perceived service quality and satisfaction, Journal of retailing, Vol. 722, pp. 201-14.

Sweeney, J.C., Soutar, G.N. and Johnson, L.W. (1997). Retail service quality and perceived value, Journal of Consumer Services, Vol. 4 No. 1, pp. 39-48.

Teas RK (1993). Expectation, performance evaluation and consumer's perceptions of quality. Journal of Marketing, Vol. 57 (4), pp.18-34.

Westbrook, R.A., and Oliver, R.L. (1981).Developing better measures of consumer satisfaction: some preliminary results. pp. 94-99 In Advances in Consumer Research, Vol. 8. Ed. Kent Monroe.

Yavas, U., Benkenstein, M. and Stuhldreier, U. (2004). Relationship between service quality and behavioural outcomes: A study of private bank customers in Germany. The International Journal of Bank Marketing, 22(2/3), 144-157.

Bennett, S. McRobb, S. Farmer, R. (1999) Object-Oriented Systems Analysis and Design, McGraw-Hill.

Burford, B. C. and Barber, C. (1993) ,A user-centered evaluation of a simulated adaptive autoteller, in S. A. Roberston (ed.) Contemporary Ergonomics, London, UK: Taylor and Francis Ltd, 117-122.

Dix, A.J. Finlay, J.E. Abowd, G. D. Beale, R. (1998) Human-Computer Interaction Second Edition, Prentice Hall Europe.

El-Haddad, A. B. and Almahmeed, M. A., (1992), ATM banking behaviour in Kuwait: a consumer survey, International Journal of Bank Marketing, 10, 25 - 32.

Hatta, K. and Liyama, Y., (1991), Ergonomic study of automatic teller machine operability. International Journal of Human Computer Interaction, 3, 295-309.

Heathcote, P.M. (2000) A-Level Computing 4th Edition, Payne-Gallway Publishers Ltd.

Miller, A.W, 2006 [Online], Who invented the ATM machine? Automated Teller History, Available at http://www.atmmachine.com/atm-inventor.html [Accessed 20/11/2006].

Hone, K. S., Graham, R., Maguire, M. C., Baber, C. and Johnson, G. I. (1998), Speech technology for automatic teller machines: an investigation of user attitude and performance. Ergonomics, 41, 962-981.

Mankze, J. M., Egan, D. H., Felix, D. and Krueger, H. (1998), What makes an automated teller usable by blind users? Ergonomics, 41, 982-999.

Mead, S., Walker, N. and Cabrera, E. F. (1996), Training older adults to use automatic teller machines. Human Factors, 38, 425 - 433.

Pepermans, R., Verleye, G. and Van Capellen S. (1996), `Wallbanking', innovativeness and computer attitudes: 25 - 40 year-old ATM-users on the spot. Journal of Economic Psychology, 17, 731 - 748.

Preece, J. (1994) Human-Computer Interaction, Addison-Wesley Publishing Company.

Pressman, R. (1997) Software Engineering, A Practitioner's Approach 4th Edition, McGraw-Hill.

Rogers, W. A. and Fisk, A. D. (1997), ATM design and training issues. Ergonomics in Design, January 1997, pp. 4 - 9; http://hfes.org/Publications/TOC/EID-93-97.html

Rogers, W. A., Cabrera, E. F., Walker, N., Gilbert, D. K. and Fisk, A. D. (1996), A survey of automatic teller machine usage across the adult lifespan. Human Factors, 38, 156 -166.

Rogers, W. A., Gilbert, D. K. and Cabrera, E. F. (1997), An analysis of automatic teller machine usage by older adults: a structured interview approach. Applied Ergonomics, 28, 173 -180.

Rugimbana, R. and Iversen P. 1994, Perceived attributes of ATMs and their marketing implications. International Journal of Bank Marketing, 12, 30 - 35.

Sommerville, I. (2004) Software Engineering Edition 7, Pearson Education Limited.

The RadioHead, (2005) [Online], The ATM Experience,

Available at http://theradiohead.blogspot.com/2005/02/atm-experience.html [Accessed 25/11/2006].

Van Vlient, H. (2000) Software Engineering, Principles and Practice, 2nd Edition, Wiley & Sons.

S. Kanwal, N.A. Zafar, Formal model of automatic teller machine system using Z notation, International conference on Emerging technologies (ICET), Islamabad, 2007, pp. 131–136.

Kurita, S.; Komoriya, K.; Uda, R., "Privacy Protection on Transfer System of Automated Teller Machine from Brute Force Attack," Advanced Information Networking and Applications Workshops (WAINA), 2012 26th International Conference on , vol., no., pp.72,77, 26-29 March 2012

ATM of Banks: Fair Pricing and Enhanced Access - Draft Approach Paper, Reserve bank of India, Technical report, 2007.

Lasisi, H.; Ajisafe, A.A., "Development of stripe biometric based fingerprint authentications systems in Automated Teller Machines," Advances in Computational Tools for Engineering Applications (ACTEA), 2012 2nd International Conference on , vol., no., pp.172,175, 12-15 Dec. 2012

ATM crime: Overview of the European situation and golden rules on how to avoid it, European Network and Information Security Agency, Aug. 2009, Technical Report.

Nitin Munjal and Rajat Moona. "Secure and Cost Effective Transaction Model for Financial Services". OPNTDS-09, 2009

James J. MCAndrews "Automated Teller Machine Network Pricing – A Review of the Literature" Review of Network Economics Vol.2, Issue 2 – June 2003

A. Gaurav, A. Sharma, V. Gelara, and R. Moona. "Using Personal Electronic Device for Authentication-Based Service Access". In Communications,2008, pages 5930–5934. ICC"08, IEEE International Conference, May 2008

Wang Y., An operational semantics of real time process algebra (RTPA), International Journal of cognitive informatics and natural intelligence, p.p. 71-89, July-Sept 2008

J. Gao, J. Cai, K. Patel, and S.Shim: (2005), Wireless Payment, Proceedings of the Second International Conference on Embedded Software and Systems (ICESS05), China, pp. 367-374, .December 2005.

S. Kungpisdan, B. Srinivasan and P.D. Le: (2004), A Secure Account-Based Mobile Payment Protocol, Proceedings of the International Conference on Information Technology: Coding and Computing, IEEE CS press, Las Vegas USA, volume 1, pp. 35-39. April 2004.

Y.B. Lin, M.F. Chang, H. C.H. Rao: (2000), Mobile prepaid phone services, IEEE Personal Communications, vol. 7, pp. 6-14, June 2000.

A. Fourati, H.K.B. Ayed, F. Kamoun, A. Benzekri: (2002), A SET Based Approach to Secure the Payment in Mobile Commerce, In Proceedings of 27th Annual IEEE Conference on Local Computer Networks, Florida, pp. 136 - 140, November 2002.

W. Adi, A. Mabrouk, A. Al-Qayedi, A. Zahro: (2004), Combined Web/Mobile Authentication for Secure Web Access Control, Wireless communications and Networking conference, IEEE Communications Society, pp. 677- 681. March 2004.

MasterCard Inc.: (1997), SET Secure Electronic Transaction Specification, Book 1: Business Description, MasterCard Inc., May 1997, http://www.win.tue.nl

A. Qadrei, S. Habib, Allocation of Heterogeneous Banks' Automated Teller Machines, First International Conference on Intensive Applications and Services, Valencia, 2009, pp. 16–21.

Tsutomu Matsumoto,Hiroyuki Matsumoto,Koji Yamada,Satoshi Hoshino," Impact of Artificial "Gummy Fingers" on Fingerprint Systems," Optical Security and Counterfeit Deterrence Techniques IV,Rudolf L. van Renesse, Editor,Proceedings of SPIE Vol.4677,pp.275-289,SPIE — The International Society for Optical Engineering,2002.

J. Hall, S. Kilbank, M. Barbeau, E. Kranakis: (2001), WPP: A Secure Payment Protocol for Supporting Credit- and Debit-Card Transactions over Wireless Networks, IEEE International Conference on Telecommunications (ICT), Bucharest, Romania, Volume 1, June 2001.

V. Pasupathinathan, J. Pieprzyk, H. Wang and J.Y. Cho: (2006), Formal Analysis of Card-based Payment Systems in Mobile devices, Fourth Australasian Information Security Workshop, Conferences in Research and Practice in Information Technology, Vol.54, pp. 213-220, January 2006.

Ayu Tiwari, Sudip Sanyal, Ajith Abraham, Sugata Sanyal and Svein Knapskog: (2007) A Multifactor Security Protocol For Wireless Payment-Secure Web

Authentication using Mobile Devices, IADIS International Conference, Applied Computing 2007, Salamanca, Spain, pp. 160-167, February 2007

Mitzenmacher M., Upfal E.: (2005), Probability and Computing: Randomized Algorithms and Probabilistic Analysis, Cambridge University Press, New York, NY.,2005.

Soriano M. and Ponce D.: (2002), A Security and Usability Proposal for Mobile Electronic Commerce, IEEE Communication Magazines, Volume 40, Issue 8, pp. 62- 67, August 2002

William Stallings: (2003) Cryptography and Network Security Third edition, Pearson Education, 2003

Index